IEE POWER SERIES 16

Series Editors: Professor A. T. Johns
J. R. Platts
Dr. D. Aubrey

ELECTRICITY ECONOMICS AND PLANNING

Other volumes in this series:

Volume 1	**Power circuits breaker theory and design** C. H. Flurscheim (Editor)
Volume 2	**Electric fuses** A. Wright and P. G. Newbery
Volume 3	**Z-transform electromagnetic transient analysis in high-voltage networks** W. Derek Humpage
Volume 4	**Industrial microwave heating** A. C. Metaxas and R. J. Meredith
Volume 5	**Power system economics** T. W. Berrie
Volume 6	**High voltage direct current transmission** J. Arrillaga
Volume 7	**Insulators for high voltages** J. S. T. Looms
Volume 8	**Variable frequency AC motor drive systems** D. Finney
Volume 9	**Electricity distribution network design** E. Lakervi and E. J. Holmes
Volume 10	**SF$_6$ Switchgear** H. M. Ryan and G. R. Jones
Volume 11	**Conduction and induction heating** E. J. Davies
Volume 12	**Overvoltage protection of low-voltage systems** P. Hasse
Volume 13	**Statistical techniques for high-voltage engineering** W. Hauschild and W. Mosch
Volume 14	**Uninterruptible power supplies** J. D. St. Aubyn and J. Platts (Editors)
Volume 15	**Principles of digital protection** A. T. Johns and S. K. Salmon

ELECTRICITY ECONOMICS AND PLANNING

T. W. Berrie

Peter Peregrinus Ltd. on behalf of the Institution of Electrical Engineers

Published by: Peter Peregrinus Ltd., London, United Kingdom

© 1992: Peter Peregrinus Ltd.

Apart from any fair dealing for the purposes of research or private study, or criticism or review, as permitted under the Copyright, Designs and Patents Act, 1988, this publication may be reproduced, stored or transmitted, in any forms or by any means, only with the prior permission in writing of the publishers, or in the case of reprographic reproduction in accordance with the terms of licences issued by the Copyright Licensing Agency. Inquiries concerning reproduction outside those terms should be sent to the publishers at the undermentioned address:

Peter Peregrinus Ltd.,
Michael Faraday House,
Six Hills Way, Stevenage,
Herts. SG1 2AY, United Kingdom

While the author and the publishers believe that the information and guidance given in this work is correct, all parties must rely upon their own skill and judgment when making use of it. Neither the author nor the publishers assume any liability to anyone for any loss or damage caused by any error or omission in the work, whether such error or omission is the result of negligence or any other cause. Any and all such liability is disclaimed.

The moral right of the author to be identified as author of this work has been asserted by him/her in accordance with the Copyright, Designs and Patents Act 1988.

British Library Cataloguing in Publication Data

A CIP catalogue record for this book
is available from the British Library

ISBN 0 86341 282 3

Printed in England by Short Run Press Ltd., Exeter

Contents

	Page
Foreword by the President of the IEE 1990/91	xi
Prologue—Purpose of the book and how to use it	xiii
Executive summary	xv
1 World outlook	xv
2 Electricity sector	xv
3A Demand assessment	xvii
3B Demand management and reliability	xviii
4A Efficiency	xix
4B Conservation	xxi
4C Environmental maintenance	xxi
5A Private versus public	xxii
5B Regulation	xxiv
5C Financing	xxv
6.1 Prescribed pricing	xxvi
6.2–	
6.6 Dynamic pricing	xxvi
7 Development programmes	xxvii
8 Developing countries	xxix
1 World outlook	**1**
1.1 Introduction	1
1.2 The world economy	1
1.3 World energy	5
1.3.1 Energy and growth	5
1.3.2 Developing countries	7
1.4 World energy sources	8
1.4.1 Resources	8
1.4.2 Non-renewable indigenous	9
1.4.3 Renewable indigenous	11
1.4.4 Internationals	12
1.5 World fuel demands	15
1.5.1 Oil	15
1.5.2 Natural gas	19
1.5.3 Coal	24
1.6 Energy management, efficiency, conservation and the environment	28
1.6.1 Energy supply–demand management	28
1.6.2 Environmental economics	30
1.7 Future energy supply–demand	33
1.8 Commentary summary	35
1.9 References and further reading	36

2 Electricity sector assessment — 37
- 2.1 Energy planning — 37
  - 2.1.1 Background — 37
  - 2.1.2 Improving efficiency — 37
  - 2.1.3 Improving investment — 38
- 2.2 National economic position — 39
- 2.3 National energy and fuels position — 40
  - 2.3.1 Introduction — 40
  - 2.3.2 Energy assessment studies — 40
  - 2.3.3 Improving energy efficiency — 40
- 2.4 Ownership patterns — 41
- 2.5 National electricity position — 41
- 2.6 Electricity markets position — 43
  - 2.6.1 Introduction — 43
  - 2.6.2 Markets — 43
  - 2.6.3 Market assessments — 43
  - 2.6.4 Consumer reaction — 44
  - 2.6.5 Competition from other fuels — 44
- 2.7 Individual utility position — 45
- 2.8 Pricing assessments — 45
- 2.9 Future market shares — 46
- 2.10 Regulators — 46
- 2.11 Commentary summary — 47
- 2.12 References and further reading — 49

3 Demand forecasting, management and reliability — 50
3A DEMAND FORECASTING — 50
- 3A.1 Traditional forecasting — 50
  - 3A.1.1 Introduction — 50
  - 3A.1.2 National economy and energy — 51
  - 3A.1.3 National electricity — 54
  - 3A.1.4 Market shares — 55
  - 3A.1.5 Commentary summary — 56
3B DEMAND MANAGEMENT AND RELIABILITY — 57
- 3B.1 Introduction — 57
- 3B.2 DM by apparatus — 57
  - 3B.2.1 Demand curtailment — 57
  - 3B.2.2 DM effects — 58
  - 3B.2.3 DM communication — 58
  - 3B.2.4 Demand controllers — 61
  - 3B.2.5 Available options — 61
- 3B.3 DM by contract — 62
- 3B.4 DM by pricing — 62
  - 3B.4.1 Use of pricing — 62
  - 3B.4.2 Price incentives — 63
- 3B.5 Consumer response — 63
  - 3B.5.1 Controlled utilisation — 63
  - 3B.5.2 Case study 1 — 66
  - 3B.5.3 Case study 2 — 72
- 3B.6 Reliability — 74
  - 3B.6.1 1990s approach — 74
  - 3B.6.2 Case study I — 75
  - 3B.6.3 Case study II — 78
- 3B.7 Commentary summary — 81

Contents vii

3B.8	References and further reading	84

4 Efficiency, conservation and the environment — 86

4A EFFICIENCY — 86
- 4A.1 Introduction — 86
  - 4A.1.1 Acceptable losses — 86
  - 4A.1.2 Studies — 89
- 4A.2 Improving production efficiency — 89
  - 4A.2.1 Thermodynamics — 89
  - 4A.2.2 Improving efficiency — 90
  - 4A.2.3 Station losses — 91
- 4A.3 Combined heat and power — 91
  - 4A.3.1 Introduction — 91
  - 4A.3.2 Industrial CHP — 91
  - 4A.3.3 An example — 93
- 4A.4 Improving network efficiency — 95
- 4A.5 Improving utilisation — 96
- 4A.6 Financing improvements — 97
- 4A.7 Commentary summary — 98

4B CONSERVATION — 99
- 4B.1 Introduction — 99
- 4B.2 National policy — 99
  - 4B.2.1 Energy shortage — 99
  - 4B.2.2 Electricity elasticity — 100
- 4B.3 Electricity conservation — 101
  - 4B.3.1 Meaning — 101
  - 4B.3.2 Market conservation — 101
- 4B.4 Who pays? — 103
  - 4B.4.1 Utility role — 103
  - 4B.4.2 Residential conservation — 104
- 4B.5 Future outlook — 105
- 4B.6 Commentary summary — 105

4C ENVIRONMENTAL MAINTENANCE — 106
- 4C.1 Introduction — 106
- 4C.2 National and energy policy — 107
- 4C.3 Electricity policy — 108
  - 4C.3.1 Environmental effects — 108
  - 4C.3.2 Remedies — 109
  - 4C.3.3 Utilities — 110
- 4C.4 Economics and who pays — 110
  - 4C.4.1 Markets — 110
  - 4C.4.2 Remedies — 111
  - 4C.4.3 Nuclear — 111
- 4C.5 Future outlook — 113
  - 4C.5.1 Reducing emissions — 113
  - 4C.5.2 Needs — 115
- 4C.6 Commentary summary — 115
  - 4C.6.1 Introduction — 115
  - 4C.6.2 Issues — 116
  - References and further reading — 117

5 Public versus private — 119

5A OWNERSHIP — 119
- 5A.1 Introduction — 119
- 5A.2 Pros and cons — 119
  - 5A.2.1 Against private ownership — 119
  - 5A.2.2 Improving utilities — 120
  - 5A.2.3 Privatisation case — 120
- 5A.3 Theory of the firm — 122
  - 5A.3.1 Step by step — 122
  - 5A.3.2 Priorities — 122
  - 5A.3.3 Dos and don'ts — 123
- 5A.4 National policy — 123
- 5A.5 Electricity policy — 127
- 5A.6 Utilities policy — 128
  - 5A.6.1 Criticisms — 128
  - 5A.6.2 Local ownership — 128
  - 5A.6.3 New entrants — 130
- 5A.7 Transmission specialism — 131
  - 5A.7.1 Introduction — 131
  - 5A.7.2 Common carrier grid — 131
  - 5A.7.3 Marketmaker grid — 131
  - 5A.7.4 Purchasing agent grid — 132
  - 5A.7.5 Distributor-owned grid — 132
  - 5A.7.6 Generator-owned grid — 133
- 5A.8 Economics — 133
  - 5A.8.1 Long-term contracts — 133
  - 5A.8.2 Pooling and settlement — 134
  - 5A.8.3 Research needed — 135
- 5A.9 Commentary summary — 136
  - 5A.9.1 Lessons — 136
  - 5A.9.2 Independent generation — 136
  - 5A.9.3 Developing countries — 137

5B REGULATION — 137
- 5B.1 Introduction — 137
- 5B.2 Types — 138
  - 5B.2.1 Heavy or light — 138
  - 5B.2.2 Examples — 139
- 5B.3 National policy — 140
- 5B.4 Legal aspects — 143
  - 5B.4.1 Transparency — 143
  - 5B.4.2 Establishing monitoring — 144
  - 5B.4.3 Legal prerequisites — 144
  - 5B.4.4 Location — 145
- 5B.5 Commentary summary — 145

5C FINANCING — 146
- 5C.1 Introduction — 146
- 5C.2 Financing types — 146
  - 5C.2.1 Government guarantees — 146
  - 5C.2.2 Foreign monies — 146
  - 5C.2.3 Special role of MIGA — 147
- 5C.3 Financial engineering — 147
- 5C.4 Financial returns — 148
- 5C.5 Commentary summary — 149

		Contents	ix

References and further reading — 150

6 Pricing — 151
6.1 Prescribed pricing — 151
6.1.1 Introduction — 151
6.1.2 Accounting cost pricing — 151
6.1.3 Marginal cost pricing — 154
6.1.4 Peak and off-peak — 156
6.1.5 Time-of-use pricing — 160
6.1.6 Retail tariffs and consumer response — 162
6.1.7 Commentary summary — 164
6.2 Improving markets — 165
6.2.1 Sector changes — 165
6.2.2 Criteria — 165
6.2.3 Market transactions — 166
6.3 Dynamic, spot and real-time — 168
6.3.1 Differences — 168
6.3.2 Step 1: Define hourly spot prices — 170
6.3.3 Step 2: Specify utility–consumer transactions — 170
6.3.4 Step 3: Implement marketplace — 171
6.4 Spot market implementation — 172
6.4.1 Developed country — 172
6.4.2 Large industrial — 173
6.4.3 Residential — 174
6.4.4 Developing countries — 174
6.5 Regulatory aspects — 175
6.5.1 Towards deregulation — 175
6.5.2 Regulated marketmaker — 175
6.5.3 Generators — 176
6.5.4 Consumers and others — 177
6.5.5 Balancing supply–demand — 177
6.5.6 Long-term planning — 178
6.6 Commentary summary — 179
6.7 References and further reading — 180

7 Development programmes — 181
7.1 Introduction — 181
7.2 Generation programmes — 182
7.2.1 Plant types — 182
7.2.2 National electricity — 183
7.2.3 Wholesale market — 184
7.2.4 Optimum programmes — 184
7.2.5 Finance — 186
7.2.6 Commentary summary — 189
7.3 Transmission programmes — 189
7.3.1 Formulation — 189
7.3.2 Options — 191
7.3.3 Rules — 191
7.3.4 Wheeling — 192
7.3.5 Regulation — 194
7.3.6 Commentary summary — 195

x *Contents*

	7.4	Distribution programmes	196
		7.4.1 Types	196
		7.4.2 Reliability	198
		7.4.3 Distribution markets	198
		7.4.4 Optimum development	199
		7.4.5 Commentary summary	201
	7.5	Utilisation programmes	202
		7.5.1 Introduction	202
		7.5.2 Reliability	202
		7.5.3 Existing utilisation plant	203
		7.5.4 Expansion of the firm	204
		7.5.5 Pricing	204
		7.5.6 Commentary summary	205
	7.6	References and further reading	205
7A APPENDIX TECHNOLOGY TRANSFER			206
	7A.1	Introduction	206
	7A.2	Transfer process	206
	7A.3	Appropriate technology	207
	7A.4	Innovation	207
	7A.5	Scale economies	208
	7A.6	Training	209
	7A.7	Patents	211
	7A.8	The future	211
	7A.9	Further reading	212
8	**Developing countries**		**213**
	8.1	Introduction	213
		8.1.1 Special features	213
		8.1.2 Economic growth	213
		8.1.3 Externalities	221
	8.2	Energy and development	234
		8.2.1 Current position	234
		8.2.2 Case study	234
		8.2.3 Energy elasticity	240
		8.2.4 Lessons	242
	8.3	Electricity and development	249
		8.3.1 Electricity's importance	249
		8.3.2 Electricity's position	254
		8.3.3 Prospects	255
	8.4	Government role	258
		8.4.1 Required roles	258
		8.4.2 Sector adjustment	258
	8.5	Special issues and options	259
		8.5.1 Pricing	259
		8.5.2 Efficiency, conservation and the environment	259
	8.6	Special financing	260
		8.6.1 Financial engineering	260
		8.6.2 Investments	263
		8.6.3 Financing sources	264
		8.6.4 Guarantees	265
	8.7	Commentary summary	266
	8.8	References and further reading	268
Epilogue—Where to go from here			**269**
Index			**271**

Foreword

I have known Tom Berrie and his work for many years. He portrays the type of engineer, much needed but not often found, who has absorbed the discipline of economics and can hold his own in that discipline as well as being acknowledged as a fully qualified engineer. Tom's first book on electricity economics entitled *Power system economics* was published by the Institution of Electrical Engineers in 1983 and has proved very useful to a wide variety of people. However, times have changed since then; many principles and methods described in that first book, that seemed well founded then, have either been largely replaced or at least drastically modified. Now, like all utilities, electricity must face up to competition, deregulation, privatisation, spot pricing, demand management, conservation, efficiency and environmental maintenance. New principles and methods are needed to cope with these changes. These new principles are described in this book, together with those traditional methods still applicable, such as load forecasting, prescribed pricing and many aspects of plant selection and reliability. It is interesting to read this book in conjunction with Tom's first and to note the socioeconomic developments that have occurred in less than ten years.

The Institution sees it has an important role in keeping practitioners up to date in their profession, and is publishing this new edition in the belief that engineers will find much of use.

Dr David A. Jones
President 1990–91

Prologue

Since the author published his book on Power System Economics in 1983 fundamental changes have taken place worldwide in electricity economics and planning. The methods developed in the 1960s and 1970s, outlined in that book, emphasised national economic development programmes and prescribed pricing from such programmes, optimising over the long-term. These methods were based upon long-term demand forecasts, elaborate probabilistic risk analysis for reliability, a discounted cash flow approach to project costs and benefits, and long-run marginal cost pricing. Almost everywhere, in developed and developing countries alike, this approach is giving way in planning emphasis to methods based on short-term, short payback development programmes, together with actual time-of-use, dynamic and 'spot' pricing, possibly set in real time. Furthermore, much greater emphasis is given today in electricity supply to the following: consumer response; private capital; private utilities; demand management; energy efficiency; conservation, and environmental maintenance.

The new book makes full use of the subject matter of the previous book, and takes into account the change of emphasis described above. It also introduces the issues and options likely to arise in power system economics and planning in the late 1990s and beyond.

Reading the book

The various groups of readers are recommended to read the following sections of the book:

1 *Decision takers, senior price setters, top academics*
The executive summary, especially the appropriate sections for their special interest

2 *Decision makers, middle price setters, senior academics and researchers*
The executive summary as a whole

3 *Decision analysts, junior price setters, middle academics and researchers*
The executive summary plus first section, and commentary summaries in each chapter

4 *Specialists and junior academics and researchers, also senior students*
Each individual specialised chapter

The executive summary is meant as an outline precis of the book for readers who should be made acquainted with the material therein, but for whom the time taken to read further is unwarranted.

Chapter 1 sets the scene, giving the background to all planning in the energy sector by discussing the world economic and energy outlook for the 1990s and beyond.

Chapter 2 deals with how governments' departments, electricity utilities (public and private), cogenerators, autoproducers and large electricity consumers, should make a national electricity sector assessment, this being needed from time to time to give background to their own planning.

Chapter 3 describes how the above organisations should assess the likely demand upon them for electricity; it bears in mind today's emphasis on demand management and the connected subject of reliability.

Chapter 4 outlines methods in coping in electricity planning with the factors of improving efficiency, conservation and the environment.

Chapter 5 lists the arguments for and against public versus private utilities, funding and regulation in electricity supply, together with current trends and likely future pressures.

Chapter 6 gives an up to date picture of prescribed pricing, i.e. the traditional setting of prices well in advance of the actual time of use; it then deals with dynamic pricing, i.e. setting prices near to, or actually at, their time of use, including 'spot' electricity pricing, buying electricity forward, futures electricity markets, agents, brokers and insurers.

Chapter 7 deals with how to determine, in the world of today and tomorrow, generation, transmission, distribution and utilisation plant programmes, bearing in mind competition, marketmakers and price setters, controllable demand and the need for short-term returns; also how large, medium and even small consumers should today optimise their utilisation programme, i.e. both their electricity plant and the usage of their plant. There is an annex on technology transfer, indicating its special importance to electricity supply in, for example, developing countries.

Chapter 8 discusses the special problems of developing countries, e.g. high growth in electricity demand, shortage of foreign and local capital, few skilled managers and labour, debt service problems, often vast but untapped fuel resources, balance of payments difficulties, etc.

The Epilogue gives a look to the future in electricity supply worldwide, describing the issues likely to arise and the probable options for solving these.

Executive Summary

1 World outlook

National economies are classified as: developed; resource-rich developing; and resource-poor developing. Some have but recently industrialised. Wholesale re-classification is unlikely during the 1990s but developing countries will increase their energy role. World economic and energy growth depend markedly on population growth. In developing countries labour increases are swamped by capital requirements, although population growth may occasionally be a net contributor to economic growth. After population, the second largest unknown throughout the 1990s will be the effectiveness of private funding and ownership for public utilities. Likely world economic outlook in the early 1990s is for $2\frac{1}{2}\%$ growth at best, with uncertainties on: development growth; oil prices and availability; the EEC, CIS (formerly USSR) and Eastern Europe growth prospects; debt servicing, and relieving poverty. The likely world energy outlook depends directly on development growth. Abundant world commercial energy resources are unevenly distributed while renewables such as fuelwood are not. The world energy scene must be periodically assessed, requiring much information on the following: resource bases; exploration rates; extraction–production ratios; production economics; supply risk and political, fiscal and operational restrictions. Periodically, demands need assessing in detail for oil, gas and coal, noting the oil companies' roles, energy pricing, interfuel competition, energy in development, country dependence by fuel. World and country fuel supply–demand balances in detail must be made by all concerned with energy for the short, medium and long term.

Energy management, efficiency, conservation and the environment will play increasing roles, but developing countries have not sufficient resources to cope with these, except with outside help. Fuel prices and availabilities will be cyclic, with developing countries only having a vital role after 2010.

2 Electricity sector

With respect to *energy planning*, before 1973 infinite, cheap, abundant oil necessitated no serious national energy planning, only fuel strategies. Many countries today use less commercial and largely unplanned non-commercial energy, with disastrous effects. Today many claim that market forces replace planning, yet many countries were not able, by the late 1980s, to balance energy–fuels inputs–outputs after the oil price rises of the 1970s. Energy sector and subsector planning is here to stay in the 1990s, dealing with supplies, demands, pricing, ownership, investment, energy management, regulation, efficiency, conservation and environmental maintenance.

Concerning *energy usage*, considerable improvement in energy efficiency is possible worldwide, especially by pricing. Most oil importing countries pass price rises on to consumers, whilst oil exporting countries continue pricing domestically below border values. Any earlier moves towards border pricing reverse when oil prices fall. Improvement in energy efficiency is most rewarding for large energy consumers, e.g. electricity suppliers and transport. Yet governmental or regulatory body incentives, financial or otherwise, are usually needed, also for encouraging energy conservation. Simple remedies are often more effective than capital-intensive ones, e.g. energy plant refurbishment, not complete replacement. For medium to small consumers, direct conservation measures work best, i.e. just to achieve less energy usage.

Regarding *energy supply*, World Bank reviews of energy sector and subsector supply positions suggest developing countries' commercial energy production rising from actual 1·7 billion tonnes of oil equivalent (btoe) in 1980 to over 3 btoe mid-1990s, 33% of the increase from oil production improvements, 27% from coal improvements, 22% from natural gas improvement, 18% through hydropower, these requiring efforts and actions by developed − developing countries' partnerships. The first action is always to accelerate identification, evaluation, development, and marketing of indigenous energy.

Large *investment* is needed to achieve sufficient energy for satisfactory world economic growth, despite improvements made in demand management, efficiency, conservation, etc., a working hypothesis being ten to twenty times the gross investment needed for developing countries, the latter being known with some accuracy and consistency. This means a doubling of energy's share of total investment in GDP growth from 2% to 3% of GDP, achieved in the early 1980s, to an average 5% of GDP over the years up to 1993/95. Funding world energy is thus a major problem for the 1990s, emphasising the need for: mobilising finance from all sources; simplifying financial availability and procedures, especially cofinancing; strengthening governmental and neutral guarantees; providing finance especially for technology transfer, particularly for developing countries.

Types of energy finance are: official, i.e. from governments, applied multilaterally or bilaterally; private, i.e. from commercial banks, finance institutions, etc.; suppliers' or buyers' credits, i.e. from equipment suppliers or manufacturers; and equity, especially with private utilities. All financing must fit within national accounting planning, and the US$ 130 billion requirement for energy investment for developing countries would be much higher without this requirement, especially if most financing were on hard commercial terms. Only small energy investment for developed countries is foreign, but for developing countries this type is much larger, possibly one half, or US$ 65 billion, over the ten years to 1993/95, compared with the World Bank's US$ 25 billion for the actual investment for energy in 1982, both measured in 1982 dollars. Mobilising adequate local capital for developing countries' energy will also prove most difficult, the favourite method to date being by direct treasury transfers to subsectors, thereby destroying the utilities' autonomy and accountability, and distorting their operations. Many economists now believe alternative methods will only work if energy pricing is all at border values and if private capital is always welcome. Effective finance needs strong institutions plus adequate regulation against monopolies and monopsonies. Each subsector, e.g. electricity, must optimally fit into national economic and energy programmes.

Periodically, assessment of national electricity must be done by all utilities in order to retain a proper perspective. If electricity supply means one single government-funded, owned and regulated utility, the national electricity is part of that utility's national economic and energy surveys. With many private or public utilities for production and/or distribution a separate national electricity assessment is needed, for each such assessment starts by considering national economy and energy. Privately owned utilities may need to hire expertise to do this. A more detailed look is then needed at the national electricity sector—e.g. number and types of utility, trends towards privatisation, deregulation—sufficiently detailed for utilities to make meaningful forecasts of national electricity growth rate, funding, etc. Electricity reviews must examine current and likely markets, their efficiency and degree of competition, also how electricity prices are set, who regulates the market and how well. Each utility must then examine its own position within national electricity market shares and future prospects; utility fuels' position and prices, examining its current and future pricing; likely introduction of consumer 'clever' meters, microprocessors, microcontrollers, microcomputers, enabling demand management, time-of-use tariffs, dynamic and 'spot' pricing to be introduced, wholesale and retail, plus the expected future of competitive fuels. Electricity marketplaces will dominate the sector in the 1990s, especially when demand management plus spot pricing become normal.

The influence, jurisdiction and scope, present and future, of *regulators*, needs assessing in all electricity reviews. Regulatory bodies vary between full, direct government control over utility ownership and funding, through 'heavy' regulation with authority to send for development and financing plans, persons and papers, to 'light' regulation, keeping only a low profile and intervening for flagrant abuses, especially against competition, but relying on the market and self-regulation as much as possible. Regulators' jurisdiction and scope are important matters for electricity utilities, generation, transmission and distribution, and for 'wheeling' electricity over the mains, i.e. using the network as common carrier, also for ensuring that autogenerators and cogenerators are given full scope and a fair competitive price for their electricity.

3A Demand assessment

Traditionally, *forecasting* is at the heart of planning. Given long plant lead times and lives in electricity, forecasts are made for 5 to 25 years ahead for peak demand (kW), annual kWh, plus demand patterns. However, ex-post evaluation questions tradition and long-term forecasting is suspect for the 1990s. Traditional approaches are: (i) from electricity data, forecasting peak demand (kW), using forecasted load factors to deduce annual kWh; (ii) from electricity data, forecast annual kWh by consumer class and summate, then using forecasted load factors deduce kW peak demand; (iii) from electricity data, forecast peak kW and annual kWh separately, checking derived load factors against past experience; and (iv) from national economic data, forecast annual

GDP and national energy, breaking down the latter into gas, electricity, coal, oil, etc., and converting electricity kWh into kW using forecasted load factors.

Traditional forecasting methodologies are:

(a) Extrapolation, e.g. previous growth continuing until saturation
(b) Examining time series of kW and kWh, rereviewing fuel and plant prices, likely new industries, improved efficiency, demand management, conservation and environmental measures
(c) Synthesising with the factors behind demand, e.g. predicting kW for future economic scenarios, using correlations of electricity with population, per capita income, consumer number, type and class, sales of appliances, industrial production
(d) Market surveys, e.g. electricity consumption examined using various credible tariffs, for existing and new residential communities, industrial or commercial estates, demand estimated from consumer number, type and class.

Electricity planning is not linear but interactive; historically it starts with demand forecasting using semi-exogenous data on cost and price, finding the least-cost way of supplying the forecasted demand, costing this and then pricing those electricity supplies, deciding whether consumers will pay the long-term price, wondering whether the utility can find the capital, obtain enough annual revenue, and finally, when the answer to any of these questions is 'no', repeating the cycle as many times as necessary. Bad load forecasts are economically detrimental because of vast resource cost losses and because pricing changes are always politically difficult. Principles behind forecasting are researching historical statistics and social, economic and political pointers showing how the economy, especially the energy and electricity sectors, are developing. Data sources are government and development agencies' economic reports, their energy reports and electricity sector reviews, and interviews with people in the country and sector concerned. Derived factors such as electricity growth, sales per consumer etc., must be compared with statistics from other countries and sectors. Common pitfalls in forecasting are: (i) extrapolation of trends regardless of altering features; (ii) forecasting influenced by supply-side factors only, i.e. preconceived ideas of size and composition of the plant programmes needed; (iii) not questioning whether the economy or utility has resource costs to meet the forecast, or whether consumers are willing and able to pay the price; (iv) also excessive 'number crunching' for its own sake.

3B Demand management and reliability

Demand management (DM) ideally influences consumer demand to optimise joint supply–demand operation, efficiency and cost. It has existed in many forms from the early days of electricity. The oldest form prescribes a maximum electricity flow, above which supplies are automatically cut off. DM is important for the 1990s because, with modern techniques, it is cheaper to control certain demands than to build more generation, transmission and distribution. In the past, supply shortages meant directly controlling selected consumers to limit demand at some periods, e.g. at peak. With limited investment resources such selective load shedding was the only control. Yet it was nevertheless often cheaper and/or more effective than installing new plant. Full DM can ensure more equitable loading of thermal generators, operating them at a controlled, predictable rate and thus installing new plant only where frequent output changes incur extra costs. Until the mid-1980s, one reason given for DM was saving on electricity, it being a 'premium fuel', similar to gas, but unlike oil or coal, therefore, needing special usage considerations. Such reasoning ceased when short-term financial optimisation criteria became paramount for utilities and consumers alike.

Demand management is a vital aspect of the economics of reliability for the 1990s, plus the necessity to know consumer response to reliability standards. When evaluating DM benefits, consumer welfare, including values put on electricity quality and reliability, must be known and vice versa. Optimum economic reliability occurs when the marginal benefit of increasing reliability by an incremental amount is no greater than the marginal cost of increasing the reliability, although the consumer side of the benefits are still sometimes difficult to measure. All DM in effect reduces reliability but it also forces consumers to assign priorities to different loads. Demands not contained in DM are normally of higher priority to the consumer. Consumer priorities should similarly be given to loads using different fuels, if switching between fuels is possible, either short term or long term, depending on capital stock. All utilities should be engaged in some forms of consumer load research to be able to keep up with consumer needs and reactions.

All commercial enterprises must continually match production to individual consumer and overall market demands at any time. Electricity DM provides a prime example of this, important complex production costs influencing and also being influenced by policies for DM, matched by sophisticated DM techniques. Demand management applies to all consumer types and classes for the 1990s but, to get adequate responses will undoubtedly require consumer education, plus incentives by pricing or otherwise, e.g. improved reliability. In the 1990s, utilities increasingly using financial rather than economic criteria, will present consumers with alternative prices and benefits for differing DM schemes, letting them work out their own welfare aspiration. The three DM methods are: (a) direct consumer demand control; (b) consumer contract; and (c) price. All require devices in consumers' premises if, unlike traditional load shedding, the consumer is always to be left with some choice of supply even under adverse conditions. Electricity outage costs are now well enough known to enable them to be used in the 1990s with a reasonable degree of confidence in any cost–benefit analysis on DM and reliability.

4A Efficiency

Generation *losses* improve with thermal efficiency and with reduced station usage. For example: (i) using improved technologies, e.g. combined cycle, combined heat and power; (ii) replacing and rehabilitating old boilers; (iii) uprating old generators; (iv) using more efficient hydro; and (v) replacing old turbines. Acceptable generation losses are never more than a few per cent, and transmission plus distribution losses no more than 10%, the optimum level of the distribution losses being 5%, not 20% after allowing for theft as in some developing countries. Distribution accounts for 75% of network losses, especially in such countries. Pricing often fails to keep pace with rising supply costs (e.g. during capital or fuel cost increases) leading to reduced investment and maintenance, hence increased losses, worsened even more by low prices over-stimulating demand alongside cutbacks in (mostly) distribution expenditure. Poor distribution voltages and high losses, both highly undesirable and costly, are unfortunately not spectacular compared with blackouts caused by generation or transmission failure. New generation and transmission are very visible and attractive signs of progress; distribution and loss-reducing improvements are unglamourous and unseen. Another reason for high loss costs are fuel cost rises, although modest equipment expenditure can make marked improvements in efficiency. Rules used for trading off increased investment against reduced losses assume relative costs appropriate to the 1960s between fuel prices, which influence the cost of losses, and copper or aluminium prices, affecting investment needed to improve losses by better or more hardware. Oil prices in real terms increased five-fold between 1965 and the 1980s, whereas aluminium and copper prices remained roughly constant.

In the 1990s efficiency urgently needs addressing at all levels, national economic, energy, utility and consumer: (i) formulating rules to judge between individual improvement strategies; (ii) judging the economics of alternative schemes for improving efficiency, e.g. combined heat and power, combined cycle; (iii) adopting balanced strategies towards the different electricity network components, generation, transmission and distribution; and (iv) evaluating possibly useful decision-making techniques imported from other sectors, e.g. entrepreneurship. Today, with financial criteria paramount, it is more important than ever to include all the costs and benefits attributable to any scheme for improved efficiency, also to realise that improving efficiency is a real alternative to building a new project, similar to DM, to be judged by normal cost–benefit analysis. Pressure groups, for whatever reason, often ignore particular costs and benefits, and tend to deal within only one level in the economy. Improving efficiency depends greatly on government policies, these varying from leaving things entirely to market forces to detailed central planning. Actually there is a minimum interest level below which governments cannot go, even when acting mainly through a regulator.

Overall, efficiency improvements in generation still have a high value for the 1990s especially in developing countries which may find it difficult to obtain capital for refurbishments. Extra high voltage transmission grids hold little room for improved efficiency anywhere. Electricity transmission, distribution and utilisation have room for efficiency improvement for both developed and developing countries, coupled with improving electricity marketplaces, either directly or indirectly, and either by better information flow between electricity suppliers and consumers, and/or especially with private utilities through dynamic or 'spot' pricing. Who should pay for efficiency improvements depends upon whether they are financially viable in the short term, in which case the utility or consumer should pay, or whether they give long-term economic returns, when the governments must offer the utility or consumer incentives. Third-party finance by companies outside electricity providing capital for efficiency improvement has successfully operated for years in the USA.

4B Conservation

Conflicting views about conservation exist, particularly if objectivity is obscured. Important conservation factors include: (a) roles of commercial, traditional and renewable energy; (b) feasibility and relative economics of old versus new technologies related to relative energy prices; (c) abundance and continuity of supplies; and (d) fuels substitutability. Just reducing energy usage itself is not worthwhile, the benefits of doing so must be greater than the costs; the same may be said for conservation measures which result in higher living standards without increasing energy consumption. It is important that conservation is efficient economically because energy is not the only scarce resource in most economies. Undue concentration on saving energy can, overall, be uneconomic to the nation. Justification for any direct governmental intervention in conservation requires markets such as electricity to be defective in this matter. Even when government measures have implications for efficiency, intervention which is then appropriate is still questionable. Government's prime role is to make sure that price, allowing for conservation, reflects social and economic opportunity costs reasonably well, i.e. for most fuels, the world market price because: (i) consumers then adjust properly their consumption and conservation patterns to energy and other price change; and (ii) an economic pricing policy requires neither omniscience nor intervention in decisions on conservation and usage that a multitude of individuals have to make in a host of different situations. Even so, the above is over-simplistic. It is impossible to separate out costs and benefits from intimately interrelated sets of programmes for the electricity sector on demand management, improving energy efficiency, encouraging energy conservation, and instituting environmental maintenance.

4C Environmental maintenance

Evidence is increasing that most environmental management is not a luxury but an essential for maintaining resources upon which nations depend. Developing nations must find a growth path that markedly differs from their predecessors in this respect, and industrialised countries must curb resource use and efficiently manage waste. Environmentalism used to be limited to urban and industrial wastes but now includes global warming, ozone holes, tropical deforestation, hazardous waste movement, acid rain, soil erosion, desertification, dam siltation, etc., in the energy sector alone. These problems defy using normal cost–benefit analysis because of the difficulty of quantifying benefits, but nevertheless must be addressed. In the 1980s, spurred by critics, governments adjusted towards environmental maintenance with growing evidence and conviction that environmental degradation constitutes a threat. Measurements of natural capital depreciation can translate environmental deterioration into economics, e.g. although Indonesian GDP increased 7% each year between 1970 and 1984, true growth was 4%, allowing for such depreciation. Environmental maintenance needs integrating into economic policy making at all levels, supported by economic and financial incentives which induce environmentally sound practices in utilities and consumers. This starts with international and national environmental studies identifying basic environmental issues and options, heightening environmental awareness, delineating ways of addressing them, and achieving an approach to their solution consistent to all sectors. Government environmental policies must provide a framework for integrating environmentalism into development programmes at all economic levels. For developing countries, the World Bank and other aid agencies are assisting. Where one country's actions affect another's due to shared resources regional solutions are needed. In the case of developing countries, it is still difficult to see what difference demands for environmental maintenance will mean in practice, depending upon whether they can afford to respond. However, it is easy to leave out the many benefits of environmentalism on development especially in energy on conservation, improved technology transfer, fuel substitution, greater efficiency and demand management.

Normal methodologies for finding optimal development programmes should be used for environmental concepts, especially cost–benefits analysis and efficiency pricing, but making sure all costs and benefits are included, although this is often difficult. Sound environmental projects are not likely to make good economic returns if considered separately from the project of which they are part, e.g. a hydro power plant. Yet often much environmental improvement can be made with little extra investment to the project, and even when strict economics, properly and credibly applied, comes out negative, particular environmental actions might still be viable on grounds other than economic efficiency, e.g. for safety, security and justice.

5A Private versus public

Generation is not automatically a natural monopoly because distributors or consumers can, if allowed, acquire electricity from various sources for regular operation, using costly supplies from their own small generators for emergencies, reserves or peak. Many governments today encourage private or public electricity suppliers to purchase some electricity externally, e.g. from cogenerators, autogenerators, combined heat and power. Unlike generation, transmission and distribution are natural monopolies, one network being cheaper than two or more parallel systems, but consumers do not seem to benefit automatically. The few communities still serviced by competitive distribution do not charge more and it is not obvious why, apart from maximising profit. Costs could also be reduced by using, with safeguards, the same poles to carry rival supplies and telecommunications circuits. Furthermore, it is not logical for suppliers to make franchise agreements if retail competition must fail. Some claim that electricity is a 'prime fuel', a 'merit good' to society, therefore only optimally supplyable by publicly owned utilities, arguing that consumers have always failed to appreciate its value to either themselves or the economy, not consuming enough of the prime good, electricity, unless subsidised for this purpose. The advantages of electricity in everyday life, nationally, locally, and individually, seem appreciated by all, even in the smallest village in remote develping countries, but unless there is some intrinsic quantifiable value in cleanliness, ease of use, adaptability, these faculties are not best dealt with under normal economic analysis.

Enormous economies of scale are still claimed for using large generators and transmission circuits, e.g. unit costs of 550 MW oil-fired steam generators are 20% less than those of a 225 MW size of the same type, unit costs of 250 MW nuclear reactors are about 50% more than unit costs of 1000 MW reactors. However, there is little evidence that the effect of scale economies, needing large sizes, is sufficient argument for having public not private utilities. Again, in developing countries investment programmes are not put together to meet what is best for consumers; financing plans do not directly seek economies of scale, even when supplies are provided by publicly owned utilities. Such utilities' service standards are often poor, improvements only proving to be possible in practice by introducing local small, often privately owned generators. To sum up, normally electricity supply does not involve externalities which automatically justify public ownership: (i) it is possible to charge properly for the service and to ultimately exclude non-payers from supply; (ii) if electricity is a prime fuel, a merit good, then consumers can be made to fully realise its value; (iii) there is no natural monopoly of production; and (iv) scale economies exist, but this does not mean that generation, transmission and distribution must exclude private ownership. However, transmission and distribution are natural monopolies, important enough to warrant regulation (see later).

In the 1970s privatisation as a concept for economic change was barely acknowledged. For the 1990s the concept enjoys full recognition worldwide. All countries embarking on structural economic adjustment see privatisation as an integral part of such changes. The public interventionist development policies of the 1960s and 1970s resulted in many publicly owned utilities, now widely regarded as stumbling blocks to regaining growth in sectors such as electricity. On the other hand, taken worldwide, the record of privatisation up to the early 1990s has been mixed. Even from successful application, no clear blueprint emerges, each privatisation seemingly having its own dynamics, this arguing for a case-by-case approach. However, some lessons have been learned from privatisation to date. World Bank case studies indicate: (a) the choice of policy instruments has depended on government objectives, utility financial performance and the ability to mobilise private sector resources, particularly through domestic capital markets; (b) the most commonly used methods of privatisation involve public share offerings, private share sales, sales of government or utility assets, disaggregation of utilities, creation of a holding utility, plus subsidiaries and management or employee buyouts; (c) most popular has been private sales to single buyers; (d) successful privatisation has occurred most often in a sound economic environment, e.g. without high inflation; (e) privatisation arguments are usually couched in financial not economic or social terms, e.g. shrinking budgets, mobilising finance, improving financial managements; (f) arguing for much greater roles for the private sector is best based on providing more effective use of national resources, savings and investment allocation; and (g) many governments ignore the worth to them of the immense sums available from privatisation.

Independent generator prospects for the 1990s are: new technologies point downwards in scale, price and optimum size, away from 2000 MW towards 200–500 MW power stations, units often available off the shelf; (ii) quick installation is nearly always better than other things for market competition; (iii) new, smaller power stations are best located within load centres, giving greater possibilities for cogenerators, autogenerators, combined heat and power and waste heat generators, raising maximum thermal efficiencies from the past 37% to sometimes 75%, pointing the way to municipal generators, user–supplier consortia, etc.; (iv) independent generator suppliers will form from banks, constructors, contractors, municipalities, etc. as classic sources, even coal suppliers; (v) second tier electricity contracts tend to be fuel-supplier dominated, not from truly independent generators; (vi) the stumbling block is still generators' contracts and take-or-pay percentages in the value or shape of contracts; (vii) bank finance is not easily obtainable for anything other than a take-or-pay virtually longlife contract, hindering fair competition between generators and incoming independents; and (viii) with many sectors having surplus base load generation, getting take-or-pay contracts to satisfy bankers on mainly peak load electricity is difficult.

Privatisation achieved some limited progress in developing countries by 1990, central issues not being desirability but feasibility. Fears of outside dominance, undeveloped capital markets, inflexibility in public finance, work-force opposition, and private sectors being highly dependent on state subsidies and contracts, remain common constraints. Also, many utilities have little asset value but high liabilities so that aid donors and recipient governments see privatisation as a later stage, after economic and public sector restructuring. Again, mixed progress to date does not mean cessation of privatisation as an effective government policy option and there seems no doubt that donors and borrowers will press for further privatisation alongside budgetary constraints, which will make ownership transfers necessary features of national economic reform packages in the 1990s.

5B Regulation

Regulation is the enforcement and monitoring of operational rules set to meet defined objectives by an autonomous agency accountable to government. Governments have sought to control policies and operations of publicly or privately owned utilities, partly through decrees, laws and decisions by ministers, partly by regulatory bodies. In addition to controlling prices and restricting market access, governments have regulated by controlling borrowing and investment programmes, and restricting the autonomy of directors and managers. Regulation has been aimed at efficiency, supply safety, minimal environmental impact and many other national objectives. No general regulatory model exists. Indeed there is widespread disillusionment with regulators trying ineffectively to achieve economic efficiency, national financial and operational objectives. In some cases wholesale electricity producers have been too powerful for regulators to prevent monopoly supplies.

Past unsatisfactory performance of electricity utilities worldwide is raising doubts about traditional regulation, experience indicating that neither government-owned nor government-regulated utilities are efficient and complete deregulation may be the answer. This requires competition and governments to develop an environment that permits many generators to sell to the transmission grid. Private investors must be guaranteed reasonable investment returns, whilst asked to share the investment risks. This will enable utilities to have greater autonomy to compete with cogenerators, autogenerators, etc., and vice versa. But regulators must always monitor operating efficiency in some way. With respect to the way ahead, questions needing answers are whether regulators should: (a) examine utility costs of purchases, fuels, investment, etc.; (b) prescribe how rises in utility costs are passed to consumers; (c) monitor utility supply–demand options from time to time, always looking for the least-cost alternative? Deregulation would introduce at least some limited competition into generation in the first years after privatisation. If distributors can purchase less generator output by improving efficiency or conservation this will be at least some semblance of competition. On USA experience, utilities will not naturally seek least-cost investment programmes for anyone but themselves, and the regulator should examine the utility investment plans for this purpose in the initial planning stages. Experience in the USA also suggests that regulatory staff should not come directly from the civil service. Also, despite the bureaucratic reputation of the USA regulators, heavy regulation has advantages. It can encourage rather than stifle competition, it being easier to envisage strong competition occurring with an active regulator than with a passive one. Again, under light regulation, utilities may use their monopoly power to protect themselves from competition. Experience in the USA has demonstrated that light, hands-off regulation has continuously allowed costly inefficiencies to occur for both shareholders and consumers, whereas purposeful regulation can more easily ensure capital is used most effectively to the benefit of all.

5C Financing

Until the early 1990s developed countries contained a mixture of publicly and privately funded, and owned, utilities, self-financing ratios varying from 20% to 80%. Other funds were borrowed from central, regional and local, government and private sources, under a variety of conditions, most with lower interest rates and longer payback periods than normal commercial loans. Privately financed and owned utilities became more common in the late 1980s, brought about by efforts in both developed and developing countries to introduce the private sector into utilities, these sectors being highly capital intensive, needing too much for public finance to supply on its own. It was also believed that the private sector would automatically introduce some measure of competition into utilities and improve marketplace efficiency. Private finance, however, has much more stringent terms than public, interest rates being higher and payback periods shorter. Electricity financing requirements will remain huge from both official and private sources: official sources involving governments directly or perhaps indirectly (e.g. in guaranteeing loans); private funding from commercial banks, financing institutions, stock markets, pension funds, insurance companies, etc., all of which have their advantages and disadvantages and the art, only learned by experience, of financial engineering is to find the mixture of financing sources optimum for a particular utility or project, the short, medium and long term all being catered for. Drawing up optimum financing plans is just as important as drawing up optimum development programmes, from which they follow after making assumptions about construction costs, fuel costs, construction times, and financial sources. Financial viability of a utility, year by year, is best measured by adopting a figure(s) for the annual rate of return on assets, turnover or investment, plus some debt service measurement.

6.1 Prescribed pricing

Traditionally electricity prices have been set well in advance of usage using the historic utility accounts to recover accounting costs, discriminating perhaps between peak and off-peak by two-part tariffs. Marginal cost pricing grew in importance from the 1960s onwards. This replaces average accounting costs per kWh by the incremental costs for supplying one extra kWh at any instant in time. Around the time of the peak, such marginal costs will have a capacity (kW) component. Marginal cost pricing should always start with strict marginal costs, even though these must then be changed, partly from axioms of cost minimisations for optimum investment programming, or from marginal cost derivation, or for meter reading practicalities and financial revenue recovery considerations, before leading to tariffs. Such tariffs, prescribed some time before electricity usage, have been well tried using long-run marginal costs derived from optimum development programmes. Degrees of sophistication have varied from simple averaged single block tariffs per kWh to complex wholesale, bulk supply tariffs, with variations in both kW and kWh rates for time of electricity usage. There is much scepticism for continuing with prescribed pricing in the 1990s, many believing this pricing does not achieve its objectives.

6.2–6 Dynamic pricing

Electricity markets in the 1990s are changing, becoming more like commodity markets, and this encourages time-of-use tariffs, also dynamic or 'spot' pricing, i.e. using a price per kWh which varies continuously, certainly each hour and from node to node in the network, matching as near as possible to actual costs of generation, transmission, and distribution. This pricing is thus connected with short-run marginal costs (SMC), not the long-run marginal costs (LMC) of prescribed pricing. Thus allocating variable costs to kWh charges and capital costs to kWh charges is not part of dynamic or spot pricing. Spot pricing also applies automatically to buy-back electricity prices, where utilities buy consumer-generated energy from cogenerators, autogenerators, or CHP, but the use of consumer capacity credits, or standby or backup charges, form no part of spot pricing. Rather, at times of capacity shortage, a quality of supply premium is added to the short-run marginal cost to always keep demand within supply capacity, working in conjunction with DM schemes (Chapter 4). A spot marketplace has many benefits for producers, consumers and the economy; operating efficiency improvements, capital investments reductions, improved consumer options on supply quality or reliability, and lower electricity prices. Such a marketplace can be implemented using existing proven technologies, and its existence will stimulate further development of the technologies required, new microelectronics for communication, and also computation with control circuits, enabling further lowering of the manufacturing costs of these.

Only time will tell how consumers react to spot pricing, especially consumers currently receiving subsidies under prescribed pricing. Again, how consumers will use equipment under spot pricing is not certain. Large industrial and commercial consumers have for some time had demand management and can easily reprogramme their communications, controls, computers, to spot pricing on a continuous or 5-minute or 24-hour update. All consumers including residential, presently manually adjusting apparatus usage to time-of-use tariffs could use computers, either to teach them how to reschedule loading or reschedule apparatus automatically to price signals. Under spot pricing investment will become partly entrepreneurial. The net revenue of a generator investment is the difference between its running costs and its revenues from selling electricity at the spot price. An incremental investment should be made if the incremental investment is less than the forecasted value of these net revenues, suitably discounted. What is not clear is how uncertainty should be allowed for in this; also how the same type of calculation should be done for transmission and distribution.

A case can be made for appointing marketmakers of the spot price in a similar manner to commodity price marketmakers, but contrary views need examining. Perhaps full spot pricing should be used only when a good proportion of generation is owned by consumers, when spot pricing will automatically lead to least-cost usage of all generators in merit-order of SMC of production, but without the formal need for a central dispatch or even a wholesale marketmaker. However, the absence of at least a wholesale marketmaker may lead to collusion between large generators to force up prices and instability in price control of the joint supply–demand system. Spot pricing provides for setting up of markets with good competition. When most supplies are from generator utilities, making the national grid the wholesale marketmaker makes sense, setting a mark-up to cover transmission and distribution losses and new lines. Regional marketmakers make sense for large countries. The concept of retail marketmakers is coupled with using the transmission and the distribution systems as a common carrier and needs much further study, as do the roles, borrowed from commodity markets, of agents, brokers, and underwriters for buying electricity forward, dealing in electricity futures markets, and insuring deals.

7 Development programmes

The 1960s and 1970s saw large expanding transmission and distribution networks worldwide. In developing countries grids are still growing, whilst developed countries' grids mostly have sufficient capacity for 2000. Environmentalists were pleased that generation originally sited near load centres was closed down, instead concentrating on large, remote, efficient plants, easily accessible to primary fuels and cooling water and using optimum technologies. But, except sometimes in developing countries, privatisation, competition, demand management, improved efficiency, conservation and spot pricing has reversed the trend back towards smaller, diesel, gas turbine combined cycle, cogenerators, CHP, sited again at the load centre. Traditional methods of determining least-cost generation development programmes are still in use for long-term planning. Due to doubts expressed from ex-post evaluation, electricity planners should at least supplement traditional methods with short-term techniques, e.g. calculate first- and second-year economic returns, plus financial analysis of the cash flow of the utility over the next three to five years.

Wind, solar, tidal, geothermal, etc., generation are still more expensive than burning fuels, with likely immediate developments only in wind and solar, e.g. large solar farms on barren land. The cost of these per kWh is still much higher than that from conventional sources but with manufacturing cost reduction of solar cells, break-even costs should be reached in the late 1990s, allowing a significant contribution from solar farms. One utility in the USA expects to supply 2190 MW of firm generation from renewables by 1994, about 15% of their requirements. Wind systems in temperate climates with unreliable sunshine need large expensive devices and find difficulty in competing with conventional, carefully controlled plant. Tidal power has construction and operation problems (especially those dealing with low head and large water volume), also firm power storage.

Developed countries' cables', overhead lines' and substations' capacity will suffice to a great extent well into the 2000s. Any new plant will see large physical changes: lighter, compacter, solid-state switches containing no moving parts and more controllable; transformers with low losses which encourage the replacement of existing lossy types, the energy saved paying for the replacement; high voltage equipment fully encapsulated in cast resin or polymers or insulated with non-inflammable, non-toxic liquid; control and communication techniques with improved reliability, microprocessor based protection and controls replacing electromechanical relays. Planning least-cost transmission and distribution networks follows the same principles as for generation, from which they cannot be divorced, although short-cuts are allowable, e.g. assuming the grid to be an infinite busbar.

Efficient electricity usage leads not only to savings but also to greater usage, comfort and convenience. Up to 50% electricity used in lighting, refrigeration, house and office heating, air conditioning, etc., is savable by modern technologies and more insulation. For example, fluorescent lights with high frequency electronic ballast provide as much light as incandescent lamps nearly five times the rating, or twice the rating of conventional fluorescent tubes with choke ballast. Freezers and refrigerators can be made four to five times more efficient by adding thermal insulation and improving motor compressor design. Buildings' heat loss is halvable by thermal loft insulation, wall insulation, double glazing, etc., and air conditioned buildings can be made almost self-sufficient by recycling air through heat exchangers. Also, better control of appliances, lights and heating, especially when premises are not occupied, can substantially reduce electricity requirements using solid-state controllers and sensors. Modern processes require less electricity per unit of output than traditional, in particular production of cement and of plastics to replace steel, television and radio sets; and these trends will continue into the 2000s.

As well as utilising electricity increasingly more efficiently, residential consumers will join industrial and commercial users in saving money by controlling and adjusting electricity usage patterns. The cost of generating the next kWh is known to each generator and with modern electronics and communications this can readily be transmitted and displayed to consumers, e.g. through the time-of-use and dynamic pricing described earlier. In industry, outputs and production schedules can be arranged to minimise electricity costs for a given production target over the day or week. In commerce, controls can be exercised to spread electricity requirements for production scheduling and comfort, again to minimise costs; this is even more possible if, at high cost times, local storage or generators can be utilised. As seen earlier, in residences, computers and intelligent controllers are readily available to control lights, cookers, washing machines, kettles, refrigerators, etc., over the household wiring, acting to minimise electricity bills.

8 Developing countries

Basic contributors to electricity demand are growths in population and per capita income. Malthus did not foresee industrial revolution, technological change, family planning, agricultural 'green' revolutions, conservationalists, etc. Thus, perhaps world population will flatten out in the 2000s. For the 1990s attention must be focused increasingly on developing countries, and their social dynamics are different from those of developed nations. Europe's growth rate in the 18th and 19th century was slow, seldom over 1% per year, whereas growth rates in developing countries today typically range from 2% to 4%. From earlier experiences there is likely to be a low level income trap which only large capital formation or major technology transfer can avert. However, population changes are both a conseqeunce and a cause of economic changes and their effects are hard to trace let alone quantify; many prefer to work instead with disaggregated population data available from UN agencies. Another common approach is to deal with average growth in total population by sub-continent, at least as a starting point for cross-sectional analyses and extrapolation, to devise electricity demand, using assumed elasticities between population growth and per capita income growth and electricity growth. Development is as complex and takes few short cuts, but some lessons exist from the past, e.g. sustained income growth almost never happens without high investment and fast import growth; the ratio of developing countries' exports to imports in 1987 was almost the same as it was in 1967. Differing regional patterns are traceable as due to natural conditions, history, politics, and debt. Cross-sectional analysis and forward projections are likely to be worthwhile; the more detail available, the more worthwhile. Each developing country must amass its own information and carry out analyses with emphases only those experienced in that country will know. Individual indications for particular world regions in which the developing country falls can be obtained from the well published material from the World Bank and the UN agencies.

Some relevant, well documented indications are that the gap between the world's rich and poor widened during the 1970s and 1980s. Overall income growth in low and middle income developing economies peaked in 1982, with average person's earnings about US$ 760 a year. Some world regions were worse off in 1990 than two decades before, reference the GDP per capita in sub-Saharan Africa dropping from US$ 400 in 1968 to US$ 340 in 1988. Also, during the latter period, private consumption per capita remained unchanged in almost all developing countries, important for electricity forecasting in that private consumption measures a person's ability to purchase basics—food, shelter, rent, health, education, consumer and other goods and services, including electricity. Those having gains in per capita private consumption tended to record gains in per capita investment. Gross domestic investment per capita for most developing countries and income groups peaked in the early 1980s, then fell, echoing the pattern of GNP per capita. Because electricity is always capital-intensive, it is vital to forecast future investment funds likely to be available to the sector, private and public, local and foreign. Electricity demand is sensitive to income level by consumer class, especially domestic and commercial. Most past and present attempts to redistribute income have had little apparent success, e.g. more than 1 billion people in developing countries live in poverty, with individual incomes below US$ 370 a year. However, important to electricity forecasting, the number of poor people will decrease if developing countries can be persuaded to pursue policies of: (i) encouraging the usage of poor labour; and (ii) providing adequate social services, especially education. The number of the world's poor will increase if there is world recession, and/or if developed and developing countries pursue only short-term financial goals.

Factors originating from outside the electricity sector became important in developed countries starting about 1975, e.g. improved efficiency, demand management, conservation, and environmental maintenance. Some argue that only industrialised countries can afford these, because they usually require additional finance to be found by developing countries, so many of which are too heavily in debt already. Agencies such as the World Bank have more recently grown fully aware of the social impacts of development and have decided to introduce into their lending a goal which contains development with nature. How much extra monies will be forthcoming for this purpose will depend on individual projects. Electricity sectors must now be expected to give full attention to these externalities. Despite the large differences, something can be learned from studying the development of industrialised countries, especially with respect to the part played by energy. During the twentieth century, electricity has made an obviously large impression on development, and this is likely to increase in the 1990s, despite the need to spend more on factors such as demand management, efficiency, conservation, and the environment.

Special roles exist for government regarding electricity in developing countries such as looking after the interest of the economy rather than the utility and the consumer, with or without a regulator, and always involving reaching compromises. Developing countries' governments must develop methodologies for optimising electricity development programmes consistent with those throughout the economy, to achieve: the best consistent pricing rationales; optimumly efficient levels and terms of reference for regulation; means to reconcile short-term financial utility objectives with long-term economic national objectives; optimum use of national resources, including manpower and skills; proper training programmes; appropriate and efficient technology transfer; and adequate rural electrification, which forms such a large component of rural energy. For too many years developing countries have suffered poor technology transfer, in electricity showing itself in the following: chasing indefinite gains in economies of scale; only installing plant with high capital to operating cost ratios; not paying enough attention to geography or skilled labour shortages, or cultural prejudices; lack of training programmes; and paying too much attention to plant successful in developed countries.

Special aspects exist for financing developing countries' electricity: obtaining the right balance of foreign to local funding; ensuring that each project has appropriate finance with respect to servicing the debt, e.g. projects with long installation times obtain long-term finance; obtaining enough total, local plus foreign, finance to ensure electricity's efficient growth and operation; setting financial targets and accounting rules for public and private utilities within national economic and fiscal policy; and ensuring debt servicing is achieved by correct pricing as well as the best balance of private to public funding. In some developing countries utilities will take the lead in all or some of these matters, in others governments or regulators will do so. Since World War II developing countries' electricity has so often been publicly owned, funded, regulated and controlled. Such countries are now following developed countries towards private funding and ownership and deregulation. Debt service in the 1980s was catastrophic and the world's financing authorities, official, public and private, will not let this happen again. Electricity, notoriously capital-intensive, will not be automatically allowed the capital quotas of the 1980s, forcing changes towards generation with lower capital to operating cost ratios, and charging realistic prices to fully cover debt servicing.

Chapter 1
World outlook

1.1 Introduction

Figure 1.1 is a world map showing three simplified categories of country economic activity, excluding newly-industrialised nations, which are shown in Figure 1.2. Developing countries' population growth leads to labour increases, unfortunately swamped by increases in other needs, e.g. capital and skilled labour. Although contrary views exist, population growth remains the first great unknown in the world outlook, the second being the effectiveness of increasing private ownership and investment [1] (Figure 1.3).

1.2 The world economy

After rapid expansion [2], world growth slowed to about 2% in 1990 from 3% in 1989, reflecting moderation in most countries and contraction in the Eastern 'Bloc'. Best expected post-1990 is $2\frac{1}{2}\%$, assuming stronger developing countries' performances. Real GDP growth slowed down in such countries from 1989 to 1990, reflecting short-term financial stringency following serious macroeconomic imbalances caused by world trade slowdowns, lower non-oil commodity prices, high interest rates, etc. Likely fluctuations in its price and availability make oil's future dependent on the Middle East and possibly some ex-Soviet republics (similarly with respect to gas). Most of the 1990/1992 Iraq/Kuwait shortfalls were offset by other OPEC members, but it is difficult to assess how long such situations could continue and how often they would be tolerable. Previous attempts limiting passing on oil price increases to consumers were often counter-productive but subsidising domestic fuels wrecks fiscal policies, unduly cushioning consumers, distorting markets, producing oil products shortage, leading to even stronger pricing pressures, greater inflation, even higher interest rates, etc. Most developed countries' fiscal positions would be worsened by a continuing long-term Middle East 'crisis' situation, making it more difficult than ever post-1990 to dramatically lower the potentially 'dangerous' fiscal deficits of the UK, USA, and the fiscal surpluses of Japan and Germany.

The medium-term outlook for world economy and energy is influenced greatly by the far-reaching changes in the Eastern 'Bloc', which had inefficient, fully centrally planned production and distribution systems, but in the 1990s is moving towards market economies, often bringing systems' slowdowns in the short term. Medium-term performance should hopefully improve rapidly mid-1990s, with market incentives built on world price trading. World prospects are

2 *World outlook*

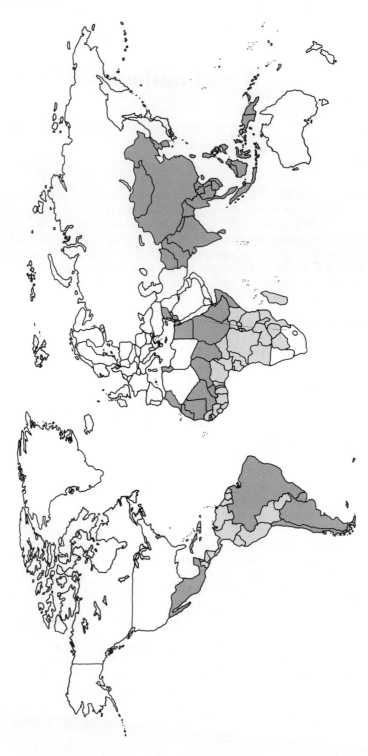

also influenced greatly by EEC market and monetary integration and whether Japan's economy remains relatively strong, periodic short-term pessimism not affecting medium-term continued, but perhaps slower, growth. Inflation will be low provided only that strong costs and price pressures are controlled. North American and UK growth may in the short-term remain sluggish, and inflation relatively high, whilst weaknesses in some financial institutions are still likely to cause concern from time to time. In the medium term inflation needs conquering for economic success, and the present serious fiscal deficits, if not strongly attacked, must affect world growth adversely as will poor resolution of world trading problems in the Uruguay and further UNCTAD 'rounds'.

Economic policies affect seriously indebted developing countries critically, being their main factor in development, medium and long term. The group will hopefully continue cautious financial policies, reducing economic distortion, encouraged by IMF/World Bank loans adjusting the structure of their economies for better efficiency. Experience to 1990 indicates the need for eliciting badly needed support of heavy external creditors, e.g. in the late 1980s, Costa Rica, Mexico, Philippines, Morocco, Venezuela, Chile, private capital playing an increasingly important role. Official creditors, e.g. the World Bank and sister organisations, must keep rescheduling debt and providing concessional development assistance; commercial financiers must demonstrate greater flexibility. Finally, for debt containment and world economic growth, industrialised countries need continued growth, with higher savings levels worldwide and open markets, all these apparently becoming increasingly difficult for the 1990s.

Global poverty remains a vital deterrent to growth, especially with respect to the energy sector and a small decrease in poverty makes for large energy demands. More than 1 billion people in developing countries live in poverty with individual incomes less than US$370 (1990 dollars) per year [3]. Poverty reduces with:

(*a*)   Industrialised nations' growing

Figure 1.1   *This world map shows how the nations of the world can be divided into three economic categories: developed (white), resource-rich less developed (grey) and resource-poor less developed (dark). For the purpose of projecting future economic development the input-output model of the world economy analyses 15 homogeneous regions that make up these categories. In 1970, the model's base year, the eight regions classified as developed, with 31% of the world population, had an average per capita income of $2,534. Three resource-rich less developed regions, with 10% of the population, had an average income of $278. Four resource-poor less developed regions, with 59% of population, had average income of $186. In the model each region is described by an input-output table, and tables are linked by a network of interregional commodity flows. (Source: Illustration by Iselin, A.D. from* Scientific American, **243** *(3), September 1980, pp. 208–209, article by Leontief, W. 'The world economy of the year 2000'. Reproduced by permission of Scientific American)*   *Illustration on previous page*

4 World outlook

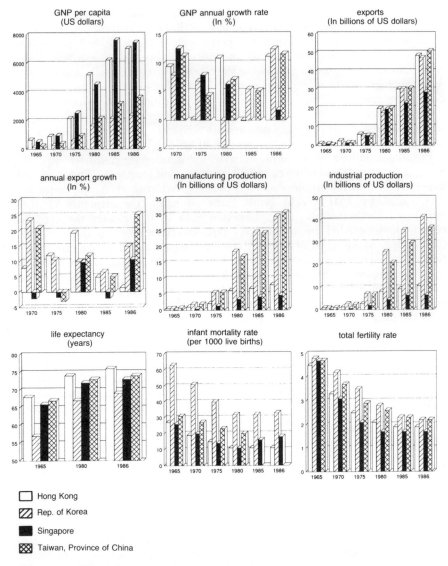

Figure 1.2 *Development of the newly industrialised economies, 1965–1986 (Source:* Finance and Development, **26**(1), *March 1989, p. 20, Washington DC, USA)*

(b) Developing nations pursuing strategies acceptable to developed countries
(c) Use of the poor's one asset, labour
(d) Provision of basic social services, especially primary education, health care and family planning.

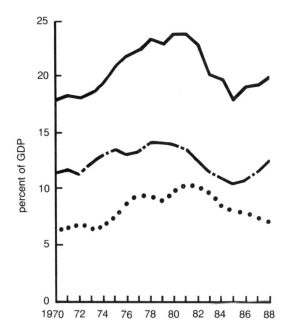

Figure 1.3 *Private investment resuming growth in developing countries (Source: Finance and Development, **27**(2), June 1980, p. 48, Washington DC, USA)*
——— total investment, – – – private investment, · · · public investment

If industrialised countries grew at 3% per annum and cash flow to developing countries increased similarly, per capita income in some developing countries would grow as fast as 5% per annum. Full use of the above factors would reduce poverty from 33% to 18% and the number of poor from 1 billion to 825 million [4].

1.3 World energy

1.3.1 *Energy and growth*

All planning must start within some logical framework (see Figure 1.4), but this is especially true for world and national energy planning. Such a framework must highlight:

(*a*) Likely periods and amounts of shortfalls between world or national energy supplies and demands
(*b*) Periods and places of abundant energy, but also likely energy supplies and demand imbalances

6 World outlook

(c) Lead time needed to be able to seek remedies before first supply and demand imbalances.

Important strengthening took place in the 1980s in world and national demand-side factors, taking better account of user response, energy management, dynamic and 'spot' pricing, fuel substitution, efficiency improvement, energy conservation, etc.

The relationship or 'elasticity' between growth in energy and growth in GDP is vitally important in the 1990s, especially for developing countries, as seen

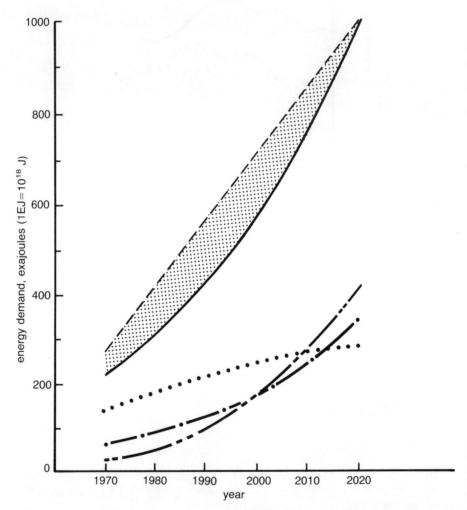

Figure 1.4 *Typical future energy demand analysis (Source:* Electronics and Power, *IEE, October 1982, p. 670. Reproduced by permission of the IEE)*
− − − potential world supply, ⎯⎯ world demand, . . . OECD demand, − · − · centrally-planned economies, − · · − developing regions

from the perusal of past elasticities. Energy planners forecast future elasticities worldwide; by continent, by nation and by region, from these obtaining meaningful, consistent estimates for future energy demands.

1.3.2 Developing countries

Developing countries are dealt with more fully in Chapter 8, but their crucial impingement on world energies is dealt with here. Before the 1970s oil price rises, commercial energy accounted for only 2% to 3% of total GDP in small developing countries with small indigenous resources, afterwards rising steeply. But commercial energy often underestimates total energy, e.g. in the late 1980s in India and Indonesia, non-commercial sources provided 50% and 60% respectively of total energy through fuelwood, dung, human and animal labour. In poor and small developing countries, energy elasticities are such that a 1% increase in per capita income requires a 1·5–3·0% increase in per capita energy. Although forecasting such elasticities is still widely used for estimating future energy demands, emphasis is today also placed on technical progress, energy conservation, demand management, improved efficiency, and environmental maintenance, all these tending to reduce energy demands for the 1990s. Previous developing countries' elasticities which varied from 1·1 for high income countries to almost 3·0 for low income countries, always tended to be well above elasticities for industrialised countries, but all such elasticities have been revised downwards after the 1970s oil price rises, suggesting that the expected economic growth of 2–3·5% for developing countries up to 1990–2000 needs an energy growth of only 2·5% p.a. However, especially in small developing countries, pursuing development for itself alone can mean choosing heavily energy-intensive projects, with demand management, efficiency, conservation and environmental maintenance applied in name only.

Post-1990, oil must still dominate world trade and is vital to all balance-of-payments positions. Fuel is one-fifth of total imports of developed countries and one-sixth for non-oil exporting developing countries. However, industrialised countries' oil flows dwarf flows to developing countries and world oil prices must remain determined by the developed world for some time to come. Many small developing countries have important, tappable non-oil energy resources (e.g. hydro, gas, or coal), but development of possibly most of these seems unlikely because enormous capital would be required. Energy forecasting in the 1990s in developing countries must be particularly flexible and be always fitted into likely, at least plausible, economic forecasts. Future values of national energy growth to economic growth elasticities are obtainable from past trends, checked against those for similar countries. Guides to future contents of national energy sectors are past and future shares of value added in energy-intensive industries and also, production per capita of these—e.g. steel, cement, glass, textiles—compared with the international or national economic average, number of motor cars, motor cycles, taxis and buses per capita; also the percentage of population living in cities.

Important general factors to take into account when forecasting national energy demands are:

(a) Developing countries' behaviour in the 1990s and beyond being more important for world energy growth than developed countries' behaviour

8 *World outlook*

(b) Developing country energy demand being volatile, with enormous potential for energy management, improved efficiency and conservation in existing and new equipment
(c) Continuing substitution away from oil
(d) Continuing worldwide large energy growth for industrial output, motor cars and urban buildings
(e) Any falling oil prices always making oil consumption attractive but rapid economic growth in developing countries meaning a great opportunity for radical change, such opportunities increasing with growth
(f) Energy pricing basically being the determining factor in energy intensity and fuel choice
(g) Nuclear being unlikely to expand in developing countries and, with difficulty at least at first, in developed countries.

1.4 World energy sources

1.4.1 *Resources*

At whatever level in the economy it is being carried out, all energy planning must start with international and national energy trends and policies, then assessing national and local energy resources however approximately, not just noting merely geological and technological evaluation of these, but wider issues. Estimates of fuel reserves and extraction costs are needed more than geological maps of fuel bearing rocks; for solar radiation, cost-effective, capturable, utilisable solar energy must be assessed rather than just radiation contours. For such assessments, it is convenient to divide energy sources into three categories: (i) non-renewable indigenous; (ii) renewable indigenous; and (iii) international.

Non-renewable indigenous includes oil, gas, coal, and nuclear. Oil includes shale and tar sands plus conventional wells, all but the latter often proving difficult and expensive to work. Gas includes gas associated with oil (petroleum gas) and gas occurring freely on its own (natural gas). Coal comprises a wide hardness range from anthracite, bituminous coal and lignite through to peat. Nuclear today means uranium. Two parameters are used worldwide for classifying potential fuel resources: (i) geological certainty of existence; and (ii) economic recovery feasibility. Teams of geologists, geophysicists and mining engineers assess basic resources. By definition, non-renewable means completely depletable fuels and these must be evaluated as such, at least identifying the above two parameters.

Renewable, indigenous resources include: solar; wind; tidal; wave; biomass; hydro; geothermal; and ocean currents. Solar includes heat and voltaic electricity; wind for water pumping and electricity; ocean systems taken strictly include wave, tidal and ocean currents. Biomass includes animal and crop residues. Hydro includes motive power or electricity. Geothermal includes heat and/or steam from underground sources, for heating and electricity. Renewables are never really limitless, many being as limited in a practical way as non-renewables, e.g. fuelwood or river systems which can support only limited usage.

	Identified resources			Undiscovered resources	
	Measured	Indicated	Inferred	Hypothetical (in known districts)	Speculative (in undiscovered districts)
Economically recoverable	Reserves		Inferred reserves		
Marginally economic	Marginal reserves		Inferred marginal reserves		
Subeconomic	Demonstrated subeconomic resources		Inferred subeconomic resources		

←Increasing degree of geological certainty —————

Figure 1.5 *Classification of energy resources (Source: 'Integrated energy planning, vol. II, energy supply' (1985), Asian and Pacific Development Centre, Kuala Lumpur, Malaysia, p. 14, fig. 10.1. Reproduced by permission of the Centre)*

Oil is an obvious supply but so also is coal, uranium, electricity and gas, by pipeline or liquid. Renewables are not tradeable internationally, except tidal, ocean currents and hydro. No international standard classifications exist for resources in the non-renewable and international categories.

The three categories of non-renewable indigenous, renewable indigenous and international are now considered in more detail.

1.4.2 Non-renewable indigenous

Non-renewable indigenous fuels information usually includes data groups as follows:

(a) Classification
(b) Exploration
(c) Extraction rates
(d) Economics of production.

With respect to (a), although no worldwide classification scheme exists, in practice the two basic parameters mentioned earlier fill the gap, namely geological certainty and economic recovery feasibility, together with some other widely used practices, e.g. those of the US or UK Geological Surveys [5] listing: (i) resources, identified and undiscovered; and (ii) reserves measured, indicated and inferred. Reserves proven, probable, or possible is also common. Figure 1.5 shows another useful, simple resources classification and many more exist.

Figure 1.6 shows a typical exploration success curve. Extensive exploration, needed to define non-renewable resources and reserves, is often unproductive and being costly needs periodic reassessment. Justifiable exploration costs are judged by estimating the value of increased reserves directly due to exploration. As exploration costs increase, exploration marginal costs also increase (Figure 1.7). Figures 1.6 and 1.7 show how, as marginal costs of exploration increase, incremental additions to reserves decrease, from which it can be judged how quickly reserves will be added, together with the exploration costs per unit of added reserves.

Optimum extraction rates depend on markets as much as on the conditions of resource deposits. Indeed, the state of coal markets is often coal's key parameter. There are maximum and minimum extraction rates dependent on resource conditions. For extraction maxima, oil is limited by well-head pressure, coal by physical rates at which current technology allows extraction, these acting as minima, possibly below economic market extraction levels.

In determining the economics of production of non-renewables, extraction costs in the form of marginal costs of production (MCP) are those for producing the next incremental unit of resource (barrel of oil, tonne of coal, etc.) i.e. the cost for getting the next unit out of the ground. Short term there are usually economies of scale, but long term the MCP will increase as extraction becomes difficult and more wells, mines, etc., are needed. Importantly, for efficiency and conservation in the 1990s, besides MCPs there will be 'economic rents', representing the scarcity values of resources, the difference between MCP and market price being additional sums the resource owner needs in order to produce a unit now instead of holding on to the resource for use later (Figures 1.8 and 1.9). In early usage, with large supplies, MCPs are low. To persuade

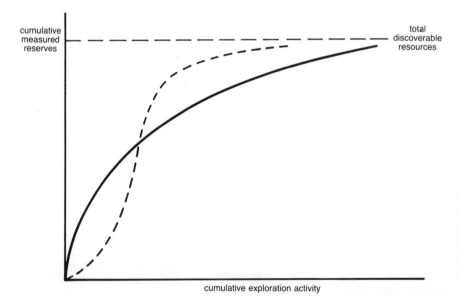

Figure 1.6 *Typical exploration success curve (the shape of the curve depends on the resource characteristics) (Source:* 'Integrated energy planning, vol. II, energy supply' *(1985), Asian and Pacific Development Centre, Kuala Lumpur, Malaysia, p. 15, fig. 10.2. Reproduced by permission of the Centre)*

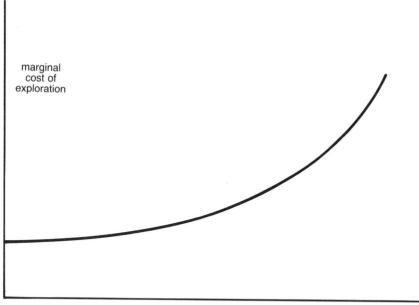

Figure 1.7 *Typical marginal cost of exploration curve (Source:* 'Integrated energy planning, vol. II, energy supply' *(1985), Asian and Pacific Development Centre, Kuala Lumpur, Malaysia, p. 15, fig. 10.3. Reproduced by permission of the Centre)*

owners to sell now rather than later, premiums are added to MCPs. As depletion takes place, MCPs increase and scarcity rent premiums become less, approaching zero. However, it must be borne in mind that very high MCPs encourage use of alternative fuels. National fuel resources are used according to different criteria—e.g. to keep domestic market prices equal to MCPs, despite scarcity, thereby stimulating artificially economic growth by artificially cheap energy, leading to rapid, long-term resource depletion.

1.4.3 *Renewable indigenous*

Similar issues to those of non-renewable indigenous resources apply to renewables but with two important differences; often no upper limit to total usable resources exists, this depending on renewable rates and utilisation linked directly with technologies available. Classification data is more difficult to obtain and divide up for renewables because energy forms and extraction processes differ greatly. A universally workable definition for size and scope of a renewables' resources base is badly needed, especially for judging exploration. As with non-renewables, the rates of extraction cannot be increased indefinitely, being dependent on: physical consideration, e.g. soil for producing biomass; technology of extraction devices, e.g. conversion efficiency of wind to electricity; market conditions, e.g. scale of domestic solar water heating and number of houses for hot water.

The economics of production for renewables differs from that for non-renewables, the MCP being a function of the available technology to convert resources into usable energy—e.g., wind into electricity—and the state of the markets. During early installation of a few units, renewables' MCPs are high. As installation spreads, MCPs fall with scale economies. Sometimes MCPs vary with installation location (Table 1.1 and Figure 1.10). Ensuring validity for the economics of renewables means using lifetime costs, i.e. using total (capital plus running) costs over the lifetime of the renewable, short-term costs consisting mostly of large capital cost ignoring the lifetime low operating costs.

1.4.4 *Internationals*

International energy is different because of technico-economic difficulties in importing supplies: renewables are not usually tradeable internationally except joint hydro schemes. Questions needing addressing in making assessments of internationals are:

(a) What sources are likely to be available and with what availability and/or standard?
(b) Export controls; using non-world-market prices, soft to hard currency conversion difficulties

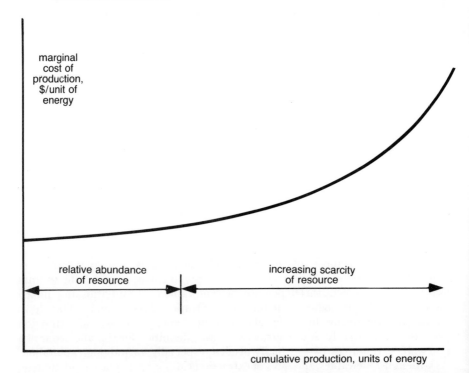

Figure 1.8 *Typical scarcity rent curve (1) (Source:* 'Integrated energy planning, vol. II, energy supply' *(1985), Asian and Pacific Development Centre, Kuala Lumpur, Malaysia, p. 17, fig. 10.4. Reproduced by permission of the Centre)*

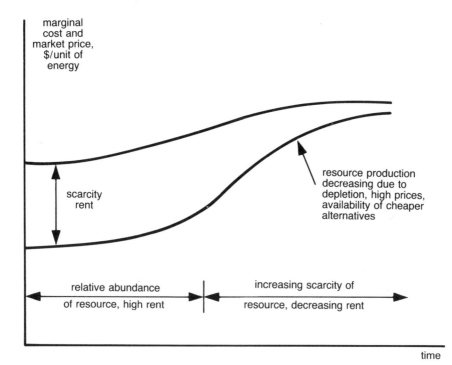

Figure 1.9 *Typical scarcity rent curve (2) (Source:* 'Integrated energy planning, vol. II, energy supply' *(1985), Asian and Pacific Development Centre, Kuala Lumpur, Malaysia, p. 17, fig. 10.5. Reproduced by permission of the Centre)*

(c) Acreage restrictions; reserving best fields for locals, differential taxes, subsidies, domestic partners requirements, joint ventures requirements
(d) Fiscal difficulties; taxation, cost recovery rules, local participation requirements, non-exportability rules, obligatory deductable costs, royalties
(e) Ring fencing; rates on profits' exclusiveness, monopolies, revenues' convertibility
(f) Cost recovery; historical and replacement cost rules, exploration cost recovery, royalties, recovery of profits, oil or gas sharing agreements
(g) Obligatory direct participation; sharing development expenses, production quotas, required rates of return on investment, interest rates, loans implicit in agreements, dealing with cash calls, joint ventures, development cost
(h) Operating difficulties; rules for operator selection, institutions needed
(i) Production difficulties; statutory and non-statutory controls, restrictions, export to local production ratios
(j) Local purchasing; rules for obtaining goods and services, non-competition arrangements
(k) Training, local labour, trainers, skills being in short supply
(l) Country versus company; sharing goals, production, finance, etc.

(m) Finance; foreign investment, capital structure regulations, profit repatriation.

The history of exploration indicates that expected after-tax returns, not price, determines supplies from new discoveries and that the oil price, etc. more than

Table 1.1 *Estimation methods for renewable resources (Source:* 'Integrated energy planning, vol. II, energy supply', *Asian and Pacific Development Centre, Kuala Lumpur, Malaysia (1985) p. 43 Table 10.15. Reproduced by permission of the Centre)*

Resource	Major parameters defining resource	Method for estimating resource base for planning

Solar energy
  Thermal
  Photovoltaic
	Incident solar radiation per unit area	Estimate either the quantity of heat provided or the amount of electricity generated. In both cases it is necessary to make some assumptions regarding technology performance

Wind	Average wind velocity	Estimate potential electrical generating capacity. Must make assumptions regarding wind turbine technology

Biomass
  Wood
  Special Crops
  Industrial waste
  Agricultural waste
  Urban waste
	Quantity of material	Estimate the heat equivalent of utilising crops. May estimate the efficiency of conversion technology (e.g. biomass plants)

Ocean systems
  Tidal
  Wave
  Ocean thermal energy conversion (OTEC)
	Height difference in tides
	Wave height and frequency
	Water temperature difference
		Estimate the potential electrical generating capacity from each system. Must make assumptions regarding the performance of each technology

Hydroelectric	Hydraulic head Flow rates	Estimate the potential electric generating capacity. Must make assumptions regarding the technology and the number of sites available
Geothermal	Temperature	Estimate either the electrical generating capacity or the quantity of heat provided. In both cases, must estimate the performance of the technology

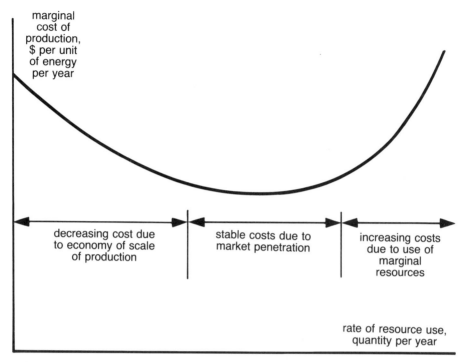

Figure 1.10 *Marginal cost of production for renewable resources (Source:* 'Integrated energy planning, vol. II, energy supply' *(1985), Asian and Pacific Development Centre, Kuala Lumpur, Malaysia, p. 42, fig. 10.6. Reproduced by permission of the Centre)*

quadrupled without collosal worldwide exploration taking place, indicating that fiscal, financial and operational burdens on MNOCs were increasing as fast as oil prices. Also inability or unwillingness of host governments to make acreage available for exploration on competitive terms helped OPEC to maintain control of world fuel markets.

1.5 World fuel demands

1.5.1 *Oil*

Some details are now given of the demand for the three primary fuels, starting with oil. Oil's success or otherwise [6] depends heavily on the supplies/demands for all primary fuels, globally, by continent, and nationally. Before the 1970s, globally, total energy grew apace with national economies and oil's share of total energy, both steadily increasing. Forecasting oil demand was easy because of the stability of fuel prices: yet oil demands in 1955, 1965 and even 1975 were under-estimated. Even after 1973, energy elasticities with respect to economic growth were still rising for developing countries, but the share of (by then) expensive oil fell. However, between the two oil price rises of 1973 and 1979, increasing world economic growth still annually added 5 million barrels per day (mbd) to OPEC's oil demand and non-OPEC oil was also becoming significantly available. After 1979, world growth faltered, leading to falling oil demand. Developed countries' energy elasticities (Figure 1.11) fell because of a shifting away from energy-intensive, heavy manufacturing towards light and service industries, and increased investment in new, more efficient equipment (Figures 1.12 and 1.13). Also there were strong changes in some consumers' response to high energy prices, including inter-fuel switching and increased competition. Non-OPEC oil's future remains uncertain within complex relationships but this oil is probably becoming more important, affecting more and more overall prices, consumer choice and total oil demand. There are many experiences to learn from. In the mid-1970s the concept of an energy gap occurring by the mid-1980s was generally accepted, but world energy growth actually became lower than economic growth, except in developing countries. Again, little understood until 1980 was how quickly non-oil sources are developable at realistic usage prices, if there is determined governmental and utilities' support. Medium and long-term energy forecasts should lie somewhere between anticipating energy or fuel gaps and just extending late 1980s energy demands, but leaning perhaps towards the latter, depending on how prices

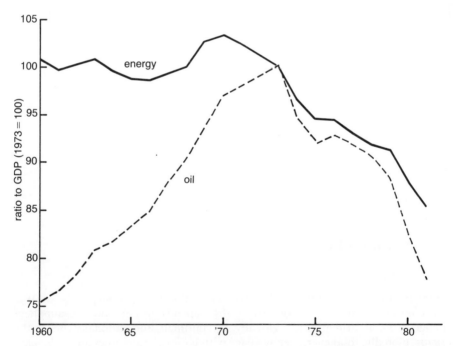

Figure 1.11 *Energy and oil intensity, 1960–1981 (Source: ROZA, A.C.A. (1985) 'The management of the oil sector'. Paper presented at the International Symposium on Energy Sector Strategy and Management, Imperial College, London. Reproduced by permission of Imperial College)*

World outlook 17

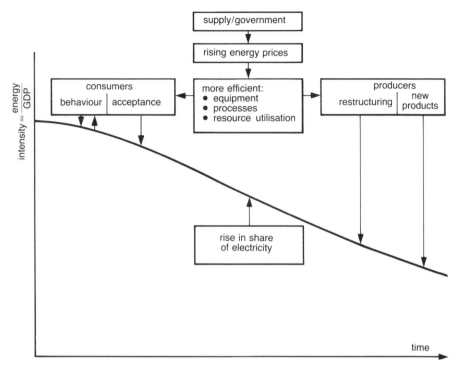

Figure 1.12 *Energy intensity in industrialised countries (Source: ROZA, A.C.A. (1985) 'The management of the oil sector'. Paper presented at the International Symposium on Energy Sector Strategy and Management, Imperial College, London. Reproduced by permission of Imperial College)*

actually move. It is always difficult to be exact about future proportions of the different fuels.

In the 1990s, despite likely continuing Middle East problems, OPEC will almost certainly dominate oil supply, but with non-OPEC oil increasingly becoming available. Future very low oil prices would mean rapidly depleted existing fields; moderate oil prices would mean some oil field recovery, plus some new fields, with non-OPEC oil unlikely to respond much. With very high prices, OPEC and non-OPEC sources would both be developed. Developed countries' energy elasticities should continue to decline with demand management, increased energy efficiency, energy restructuring of economies towards private funding and environmental maintenance. Restructuring in industry is likely to continue, including drastically changing the kinds of processes used in energy-intensive industries (e.g. iron, steel, and cement) and there will be continued growth in industries requiring only small amounts of energy (e.g. microchip and service industries) all these effects continuing, even with no further sustained oil prices rises, but because they are all economically worthwhile. Oil demand in developed countries should not grow much, ceiling levels depending on economic growth and equilibrium levels of the energy

18 *World outlook*

elasticities themselves. For most of the likely future long-term levels of world economic growth, energy demand increases will be flatter than historically, with renewables unlikely to globally play important roles before about 2020 or later, their importance depending upon the country and the location. Periodic global over-capacity, and possibly under-capacity, in oil supplies seems certain.

Inter-fuel competition will continue, especially for industry, but not as much in markets less energy-intensive. Medium- to long-term fuel choices will depend on mainly national fuel prices, these being always heavily government influenced, particularly for primary fuels, e.g. oil, gas, coal and nuclear (electricity). Oil and gas will compete mainly in domestic and commercial markets, oil continuing to globally dominate transport, certainly for the foreseeable future. Plentiful availability of all energy, short- to medium-term, puts downwards pressures on oil's share of total energy, also on pollution and safety, e.g. acid rain and nuclear wastes. It is difficult to predict how non-OPEC

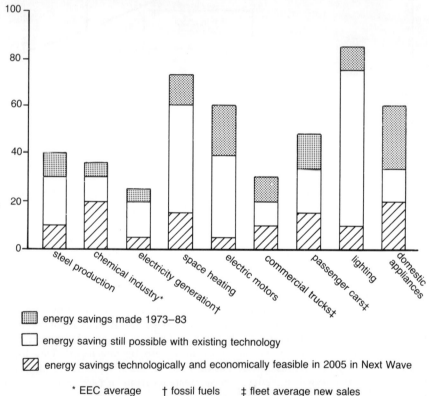

▓ energy savings made 1973–83

☐ energy saving still possible with existing technology

▨ energy savings technologically and economically feasible in 2005 in Next Wave

* EEC average   † fossil fuels   ‡ fleet average new sales

Figure 1.13 *Technological potential. This shows that only a small portion of the technological potential has been taken up. Even without allowing for technology synergy, this potential for energy saving is expected to increase (Source: ROZA, A.C.A. (1985) 'The management of the oil sector'. Paper presented at the International Symposium on Energy Sector Strategy and Management, Imperial College, London. Reproduced by permission of Imperial College)*

oil and the sources of other primary fuels will behave, especially *vis-a-vis* renewables.

Developing countries, needing considerable energy to develop, should mostly have constant or still slightly increasing energy elasticities, although newly-industrialised countries (e.g. Korea or Taiwan) will behave increasingly as developed countries. Countries which are little developed will require considerable energy, especially where most fuels have to be imported. In the extreme, if countries do not develop internal indigenous resources, or have none, their economies may not grow at all, and demands for energy will remain static, even reduce. In developing countries about 6 mbd of oil equivalent seems likely to be lost by deforestation up to 2050; however, if oil were substituted for this it would seriously affect the global oil position, as would the emergence of any large new indigenous sources or sinks, or large energy management, or improved efficiency or energy conservation programme. Inter-fuel competition is less intense in developing countries because of priorities on indigenous resources and infrastructure limitations. Dependence on imported oil must generally decrease for some countries to survive economically, regardless of medium- to long-term prices. Also, importantly, the growth in oil demand for developing countries is likely, in the late 1990s, to swamp the total decline in oil demand for developed countries, keeping the total oil demand about the same. Further discussions on developing countries' oil continue in Chapter 8.

1.5.2 *Natural gas*

Global natural gas [7] demand remains constant at about 20% of total energy, but growing in absolute terms. To meet this, proven ultimate reserves estimates grew substantially in the 1970s and early 1980s. Gas supplies are abundant for the 1990s worldwide, but are not geographically uniformly distributed, the former USSR remaining the gas 'Middle East', with not quite half the world reserves. Bulk trading is done mainly via pipelines, although oil shortages caused bottled gas use by some countries extensively, e.g. New Zealand and other countries with plentiful gas supplies. Bulk trading increased from 13% to about 14% in 1988 and 15% in 1989, long-term contracts being for two years at least. The four main gas markets are: (i) space heating; (ii) industrial steam raising; (iii) electricity generation; and (iv) feedstock for chemicals, petrochemicals and fertilizers. The last two categories are economic only at high load factors. One extreme market, Japan, uses 70% or more gas for electricity. Further gas markets depend on relative fuel prices. Historically, government pegging of gas prices *vis-a-vis* other fuels has seldom achieved stated objectives; in the 1990s more economies will leave gas prices to market forces. However, some governments still calculate marginal costs of gas production, add financial constraints plus incremental transport costs to obtain minimum gas sales prices, and possibly add some depletion rent.

Figure 1.14 illustrates the historical steady growth in gas consumption, and Figure 1.15 the primary energy gas share consistent with this trend. Figure 1.16 illustrates how the trends could give a wide practical range of estimates for world gas consumption in the year 2000, illustrating the problem of predicting from any trends. Historically, to keep the supply–demand balance, proven gas reserves, defined earlier, grew substantially, and there are still additional reserve categories (Figure 1.17). Figure 1.18, on a basis consistent with the

20 World outlook

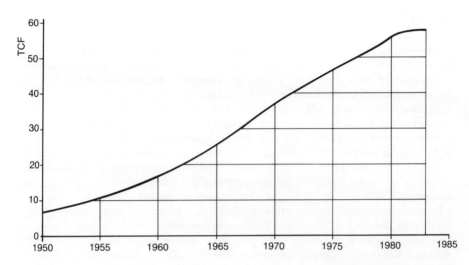

Figure 1.14 *World gas consumption 1950–1983 (Source: WESLEY, M.W. (1985) 'The management of the gas sector'. Paper presented at the International Symposium on Energy Sector Strategy and Management, Imperial College, London. Reproduced by permission of Imperial College)*

figures already quoted, illustrates the proportion of proven fossil fuel reserves; in 1960 39% of oil reserves, in 1970 45%, in 1982 86%, and in 1986 90%, at which level it has approximately remained. By the mid-1980s, on the same basis, reserves to production ratio for gas approached double that of oil but, unlike oil, there has still never been a concerted exploration for gas. Yet the resource base will not be a constraint on demand growth during the 1990s.

	UK	West Europe	USA	Japan	World
Natural gas	23	14	27	7	19
Oil	36	49	41	61	47
Coal	35	22	23	18	22
Nuclear	5	6	4	8	4
Hydro	1	9	5	6	8
Total	100%	100%	100%	100%	100%

Figure 1.15 *Primary energy consumption 1982, % market share (Source: WESLEY, M.W. (1985) 'The management of the gas sector'. Paper presented at the International Symposium on Energy Sector Strategy and Management, Imperial College, London. Reproduced by permission of Imperial College)*

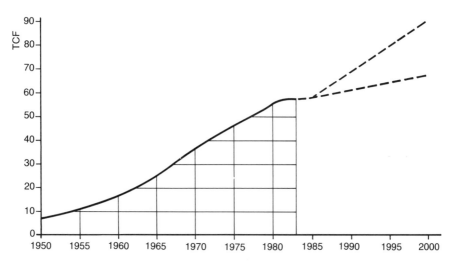

Figure 1.16 *World gas consumption, projections to 2000 (Source: WESLEY, M.W. (1985) 'The management of the gas sector'. Paper presented at the International Symposium on Energy Sector Strategy and Management, Imperial College, London. Reproduced by permission of Imperial College)*

Figures 1.19, 1.20 and 1.21 illustrate the potential for further growth locally, regionally and internationally, again on a consistent basis with the previously quoted figures.

Basically gas remains national not international, with almost 90% of consumption local. Liquified natural gas (LNG) remains only a quarter of international gas, dominated by only a few countries. Because gas is much more expensive to transport than oil, increasing with distance, gas will always be uneconomic compared with oil traded over the same distance and compared to local gas. Like electricity, gas is a clean, flexible, convenient, efficient, high-grade fuel, generally more environmentally acceptable than other primary fuels, this last becoming increasingly important in the 1990s. The four main gas markets mentioned earlier have different characteristics.

In the space heating category, residences use gas for cooking, water heating, space heating and air conditioning, loads being small and space heating seasonally temperature-sensitive. Commerce uses gas for space heating and water heating in offices, schools, shops and hospitals, loading being somewhat larger and load factors higher. In the industrial steam raising category process applications use gas's special qualities; load factors are high and loads large. Gas is also used for direct steam raising, often with other fuels, when gas supplies can be interruptable at high demand times, the overall load factor of the total gas market being then more manageable. In the electricity generation category, gas markets for electricity have large loads at high load factors, often used with other fuels and often interruptable, sometimes being sold to electricity at the 'seller's option', providing a major method of regulating total gas system load factor. In the petrochemicals, fertilisers, etc., category loads can be large and at high load factors; interruptability is not usually available, most con-

22 *World outlook*

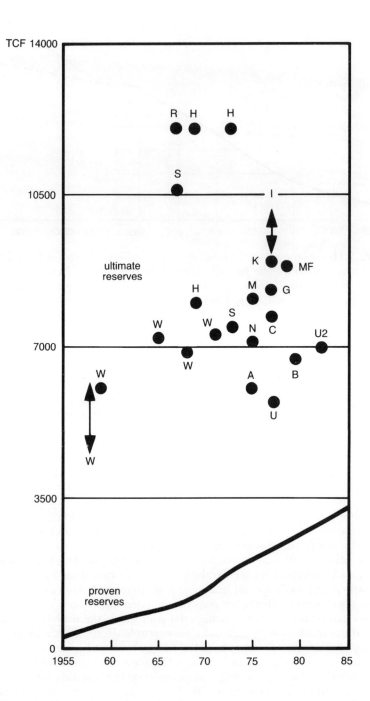

sumers' plants not having dual feedstocks, and production continuity being important. Overall, national, social, political and economic policies influence this market—e.g. Japan's concern for the environment led to large use of imported gas for electricity.

The 1990s should see increasing gas imports to major consuming nations, requiring long-term contracts between discrete buyers and sellers, often involving governments. If gas is to play an optimum role, better recognition will be needed of the points that national, continental, attainable gas prices in importing nations' markets can always be determined, and that gas is often sold by utilities at restricted prices. However, there is unlikely to ever be an international gas market price directly applicable nationally or continentally. Gas is invariably reliable, stable, and flexible, a combination of all the above factors determining gas imports' prices to any nation. Resource costs of gas are usually less than market prices, the margin depending on source and destination. Generally, trading growth depends on whether the difference in market price and resource costs meets aspirations of exporting nations or parties. In countries with significant reserves, maximization of local uses to reduce oil imports or, in oil-producing countries, to export more oil, remains desirable. However, on the one hand, gas supply systems are expensive and, on the other hand, uniform cash flows must be generated to provide proper returns on investment. Therefore, rather than as in the past, aiming at 'premium' uses for clean, convenient, flexible gas, priority should in the 1990s rather be given to acquiring high load factor/large loads, e.g. electricity generation or large industrial steam raising. However, once the gas supply infrastructure develops for these purposes, it should be economic to extend again into 'premium' markets. Developing petrochemical and similar loads for internal use, or export, means establishing expensive conversion processes alongside the fierce world competition of the 1990s, and there will be few circumstances which will be economic. Developing a gas industry with no reserves but to reduce dependence on oil and diversify energy resources for security, may be politically and socially worthwhile.

Figure 1.17 *Estimates of world gas reserves (W L.G.Weeks, R Ryman (EXXON), S Shell, N National Academie of Science (Washington), A T.D. Adams, M.A. Kirby (9th World Petroleum Congress), I Institute of Gas Technology (Chicago—January 1977), K M.A. Kloosterman (September 1977), C CME–IFP (September 1977), U International Gas Union (August 1977), H M.K. Hubbert (V.S.G.S.), M J.D. Moody, R.E. Geiger, G LNG (5 Dusseldorf, September 1977), MF Meyerhoff, (10th World Petroleum Congress, 1979), B BGR, Germany (Prepared for WEC 1980), U2 International Gas Union (Task Force II) 1982) (Source: WESLEY, M.W. (1985) 'The management of the gas sector'. Paper presented at the International Symposium on Energy Sector Strategy and Management, Imperial College, London. Reproduced by permission of Imperial College)*   *Illustration on previous page*

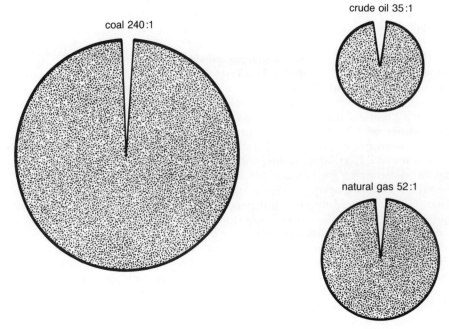

Figure 1.18 *Ratio of proven reserves to 1983 production (Source: WESLEY, M.W. (1985) 'The management of the gas sector'. Paper presented at the International Symposium on Energy Sector Strategy and Management, Imperial College, London. Reproduced by permission of Imperial College)*

1.5.3 *Coal*

Coal [8] has two natural boundary scenarios within which most outcomes must lie, depending on externalities which influence its competitive position and internalities which govern its ability to respond. Concerning externalities, a natural hierarchy of coal usage worldwide equates to coal's chemico-physical makeup, coal being intrinsically a combustion hydro-carbon fuel. Compared to oil and gas, coal has an over-abundance of carbon and too little hydrogen. Thus, to produce oil or gas from coal requires reducing carbon and increasing hydrogen, both processes being expensive. Oil and gas are always readily usable for combustion, often when coal cannot be used because of geographical, physical and environmental reasons. Indeed, for convenience and the environment, oil and gas would be used for combustion worldwide were it not for large price differentials between them and coal. Whenever and wherever oil is cheap, it will be advantageously used for combustion and there will then be greater production of light petroleum products from refineries to compete with gas. Since the 1970s most countries have tried to keep (now) expensive oil and natural gas for more 'premium' use than coal; e.g. not for combustion, coal's 'natural' job, especially in power stations. Because coal has distinct physical disadvantages of bulk, impurities, handling, etc., it performs best in world markets where its chemical and physical disadvantages can be minimised, i.e. when it can be bulk handled, again especially at power stations.

Moving up the coal usage natural hierarchy, the spectrum follows power stations, through heavy industry, steam raising and bulk heat, fitting in with middle-range industrial or commercial applications, to light commercial and residential markets; only at the very end of the hierarchy occurs conversion to gas or oil. Moving up the hierarchy, the price differential between coal, oil and gas should increase steadily in order to ensure proper consumer market penetration of coal. When coal competes with fuel oil for electricity, coal must have a price advantage of about 20%, but the largest differential needed is for coal's conversion to gas or oil, plus some petro-chemical feedstocks, where the ex-works cost per therm of the products are between two and three times the cost of the input coal. When oil was cheap, high proportions were produced for coal's main competitor fuel oil, for bulk combustion in power stations and general industry. Following the oil price rises there occurred marked fluctuations in fuel oil output and price. By far the largest market for steam coal in the 1990s will be power stations, this coal market showing major expansion since the early 1980s and likely to continue but at a reduced rate, as gas is now taking over in many privately owned power stations. Coal's long-term task is, knowing the intrinsic 'nastiness' of coal as a hydro-carbon, to prevent economic losses that would occur from using oil and gas for bulk combustion. However, the important medium-term question is how far coal can move up its own usage

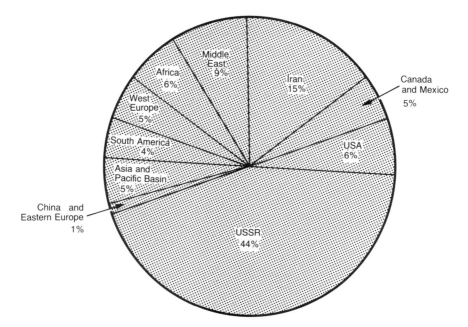

Figure 1.19 *World proven gas reserves to 1983 (Source: WESLEY, M.W. (1985) 'The management of the gas sector'. Paper presented at the International Symposium on Energy Sector Strategy and Management, Imperial College, London. Reproduced by permission of Imperial College)*

26 *World outlook*

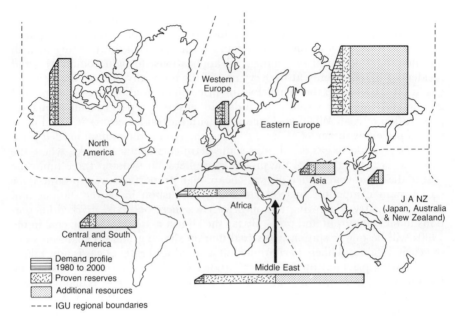

Figure 1.20   *IGU Regional classification
(Source: WESLEY, M.W. (1985)* 'The management of the gas sector'. *Paper presented at the International Symposium on Energy Sector Strategy and Management, Imperial College, London. Reproduced by permission of Imperial College)*

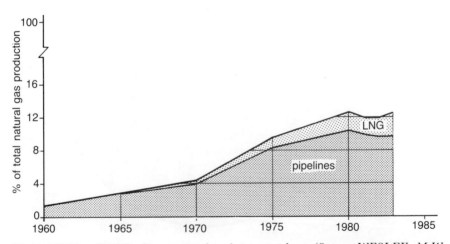

Figure 1.21   *Growth of international trade in natural gas (Source: WESLEY, M.W. (1985)* 'The management of the gas sector'. *Paper presented at the International Symposium on Energy Sector Strategy and Management, Imperial College, London. Reproduced by permission of Imperial College)*

hierarchy, diversifying out of electricity into industry or commerce; also, what kind of price advantage will then be needed? Improved coal-fired boilers, and more flexible financing will play major parts in any such changes. Thus, coal's share of world energy in the 1990s depends on whether it can expand into the more premium market and at a faster rate than its erosion from electricity and industrial usage, e.g. by gas and possibly, long-term, nuclear. Accepting this, the single, vital concern for future coal demands is the oil and gas prices, these being determined largely internationally not nationally like coal prices, and coal supply patterns, within large numbers of plausible scenarios. When estimating future national coal demands, competition must be considered, not only from other national primary fuels, but from international coal such as the long-standing international trade in coking coal and, lately, the drastic increase in international trade in power station and industrial steam coal. It is likely for some time ahead that national economic values of coal in most developed countries will be determined, directly or indirectly, from competition at the margin with internationally traded steam coal. One likely scenario is for oil prices to set effective upper limits on international coal prices, another is that the latter price will be determined mostly by interplay of supply and demand of traded coal itself, even though this will continue to represent only small proportions of total steam coals used globally. Also, because potential world coal reserves are so large and detailed knowledge of world coal supplies and demand at any particular time imperfect, the lowest cost coal reserves will not always be worked first, globally, perhaps far from it, although there is such a tendency within established coal fields. This means that there is no inevitability about rising marginal costs of coal production being forecastable worldwide for planning and setting prices. Similarly, no dominant coal cartel will form, even for steam coal, markets operating highly competitively, individual producers improving sales at the expenses of others, and buyers diversifying their supply sources. Also, foreseeably, with major markets for steam coal remaining in power stations, lead times on coal demand creation will be greater than those on coal supply creation, reinforcing the competitive nature of coal markets, meaning that markets will inherently over-, rather than under-supply. Although in some years prices have been different from those sustaining the trend in growth in international steam coal, coal prices in real terms are not expected to increase significantly up to the year 2000 at least. Because world coal reserves are extremely large, and because of their wide geographical dispersion, knowledge of production is not easily restrictable globally. Also, improvements in coal transport are continuous; coal can be landed in Europe from New South Wales at prices below best European coal.

So much for considering externalities. Concerning internalities, the starting point is in mining operations themselves. Some countries established coal mines long ago with average lives of possibly 80 years. Because at any individual colliery the best seams are worked first, natural erosion of potential capacity occurs steadily, this being compensated for by productivity increases, new technologies and new working areas. In considering the areas to develop, major geological parameters are seam thickness and degree of fault intensity, these determining mining operations' continuity. There are significant lead times to consider. In most countries, new mines will take over ten years to develop and major developments at existing mines will take five years or more. It seems unlikely that, up to 2000, there will be revolutionary changes in mining technology worldwide. However, there often exists an immense, untapped potential in the use of existing well-proven technology, evidenced by performance at the best faces which use it, producing at three or four times the rate achieved on average faces.

1.6 Energy management, efficiency, conservation and the environment

1.6.1 *Energy supply–demand management*

Chapters 4 and 5 are concerned with these subjects apropos of electricity; what is contained here is applicable to the energy sector in general within the national economy. Successful energy supply–demand management acts on all activities in the energy sector [9], thereby influencing, guiding, or controlling all energy production and consumption. Its primary purposes should be to maximise growth in net national economic welfare through the energy sector, although it must have some important effects on utilities. It begins by making energy supply–demand balances, these effectively fitting in with national economic and energy objectives. In this respect it has four components: (i) identification, verification and subsequent optimal exploitation of all energy resources; (ii) energy source planning, transformation, refining, transportation and distribution; (iii) optimising substitutions between different forms of energy; and (iv) operation and maintenance of energy supply systems. It embraces all means of influencing users' levels and patterns of energy consumption. So-called hard tools comprise: (i) direct physical control of energy supplies, up to rationing; (ii) regulations concerning energy usage; and (iii) technology options, e.g. energy-saving machinery retrofits. Hard tools are likely to 'naturally' be most effective the shorter their usage time. Soft tools, having different natures to hard ones, are more useful in the medium and long term, e.g. energy pricing, subsidies, taxes, grants, etc.

Energy efficiency improvement and conservation, to be most successful, should be part of energy management, properly definable as optimizing the joint net national economic welfare through ways which actually improve energy efficiency and reduce total energy, respectively. They both consist of much more than merely reducing the amount of energy used, or improving its usage efficiency, for these reasons alone. All workable energy management requires complete analysis of links between energy and all other sectors: inputs, e.g. capital, labour and raw materials; environmental resources, e.g. clean air, water or space; outputs, e.g. electricity, petroleum products, woodfuel, etc.; and national policies, e.g. fuel availability, prices, taxes, subsidies, etc. These must of their nature involve governments, at least in some measure. Detailed analyses are then needed for each energy subsector—electricity, coal, gas, oil—emphasising interactions among subsectors; e.g. fuel oil and coal for electricity generation, kerosene versus electricity for lighting. Each subsector must determine its own demand forecasts and optimum investment programmes.

In practice, the different levels and scopes overlap. Within certain clear limits, many energy resources substitute for others, and energy management can bring about significant shifts in demand for specific energy resources, at least long term. Similarly, deliberate aiming at energy conservation may significantly affect energy consumption patterns. Other conservation measures may substitute energy by capital or labour, e.g. replacement of pilot lights by electronic switches, recirculation of process heat in industrial plants through better engineering or lighter materials, or installation of better building insulating materials. Energy management should produce a coherent set of energy sector and/or subsector activities to best meet the needs of the many interrelated, often conflicting, national economic and national energy objectives. Only government, not utilities, consumers or market forces, can find an equitable balance by means of which all national objectives can best be achieved. In this vein, primary objectives of energy sector and subsector management must be on the lines of:

(a) Determining detailed energy requirements to achieve economic growth and economic targets
(b) Choosing the fuel mix for future energy requirements giving greatest net economic benefit
(c) Minimising unemployment and other similar socio-economic conditions
(d) Conserving energy economically, eliminating wasteful consumption
(e) Diversifying supply and reducing over-dependence on foreign sources
(f) Supplying basic energy needs of the poor
(g) Saving scarce resources, e.g. skilled labour, capital and foreign exchange
(h) Identifying specific energy demand and supply measures for economic development of special regions or sectors
(i) Raising sufficient revenues from energy sales to help finance energy sector and subsector development
(j) Ensuring price stability
(k) Preserving the environment.

The major steps involved for all forms of energy management, including efficiency improvement and conservation are:

(a) Analyse past energy supply-demand data
(b) Forecast these for future years
(c) Forecast detailed energy supply–demand balances by fuel type and consumer class
(d) Determine particular energy management policy
(e) Analyse likely policy impacts on the energy sector.

All steps must consider the short, medium and long term, although there must often be difficulty in reconciling these, one prediction with the other. It is advisable to use simple methods when this exercise is first carried out; later, with improved data, better analytical capabilities and proven methodologies, more sophisticated techniques, including computer modelling, can be used, although still, at first, with caution. Developing countries face additional constraints: severe energy market distortions due to taxes, import duties, subsidies; externalities causing actual market and financial prices to diverge substantially from economic and opportunity-cost prices. For improved economic efficiency, governments may be forced to alter, and ask energy utilities to adjust all costs and benefits associated with investment, possibly operation, towards true economic and away from straight financial values, to be modified even further by anticipated consumer reaction. Developing countries suffer from income disparities and social considerations which call for subsidised energy prices or rationing to ensure that the basic energy needs of all consumers are met.

1.6.2 Environmental economics

Environmental economics are not straightforward and must always begin with some general principles [10]:

(a) Environmental maintenance will not be easily, readily, achieved by market forces alone; although its costs are easily qualifiable, benefits are not, and assessing worthwhileness is difficult.
(b) Environmental maintenance is becoming socio-politically demanded, despite cost, requiring increasing numbers of large insurance type premiums be taken out on all developments for this purpose.
(c) Environmental regulations will always basically be socio-political acts, not results of cost–benefit analysis.
(d) Environmental concerns being international, this adds further dimensions, the number of potential conflicts of interest being much larger than the number of policy instruments; developed countries are more willing than developing to accept environmental maintenance costs, their viewpoint being different on acceptable balances between costs and benefits; market forces are even more out of place internationally than nationally.
(e) Even when environmental regulations are introduced with little economics, once in place they have significant implications on energy production and usage.
(f) The speed of introducing regulations often is in conflict with investment lead times, or lives of assets of energy suppliers and users.

For example, it seems likely that arriving at effective sulphur dioxide regulation within the EEC was, in reality, socio-political rather than by use of market forces or by cost–benefit analysis, yet these regulations are effective, e.g. resulting in modest economic costs and little distortion of markets for electricity generation fuels.

However, the greenhouse effect is likely to prove the same yet somewhat different. Given the state of knowledge, plus lack of reliable world climate models, there seems no possibility of doing cost–benefit calculations. Yet socio-political awareness is rising rapidly in the 1990s. In June 1988 the Toronto Conference on the Environment produced a 'Statement'—still receiving much attention—calling for a 20% reduction in carbon dioxide emissions by 2005, half of the reduction coming from improved energy efficiency and half from fuel-switching. The conference tended not to face up to global energy arithmetic and associated economic consequences. (The Rio Conference of 1992 did little better.)

A general idea of the importance of the greenhouse effect, etc., for the 1990s and beyond is now given. More details apropos electricity are given in Chapter 5. In the 20-year period up to the late 1980s world energy consumption rose from about 6 to about 12 billion tonnes of coal equivalent (tce), with the

Table 1.2 *Changes by fuel (Source: PARKER, M.J. (1989)* 'The economics of environmental control of energy'. *Paper given to an* ad hoc *energy forum at Surrey Economics Centre, Guildford, UK. Reproduced by permission of the author and the Centre)*

(Billion tce)	1967	1987
Coal and other solid fuel	2·35	3·41
Oil	2·53	4·20
Natural gas	1·08	2·22
Nuclear and renewables	0·39	1·33
Total world	6·35	11·16

changes by fuel illustrated in Table 1.2. OECD increases were modest, centrally planned economies (CPE) large, and developing countries rapid, from a low base; Table 1.3. These trends, although changing somewhat with socio-political pressures of the 1990s, basically will not stop. As seen earlier, world population is likely to rise by one-third by 2005, and developing countries will continue to press for more energy per capita, seeking to increase living standards. Past average annual growth rates continued to 2005 will result in energy demands and carbon outputs as shown in Table 1.4. Thus world economy and energy sectors left to themselves would increase carbon dioxide by about 75% rather than reduce it by the 20% called for by the Toronto Conference, and it is necessary to take the above figures, not early 1990s levels, as the theoretical uncontrolled bench mark. Because of world recessions, uncertainties in oil supplies and prices, such energy increases may be inherently implausible. In any case, such demands are not able to be supplied without very large rises in energy prices. Yet the difficulties of attaining energy efficiency/conservation in order to reduce world energy remain.

Recommendations which followed the Toronto Conference give some guidance (Table 1.5). Given the 1990 mix of fuels, the total world energy demand would be about 14.0 billion tce, and the absolute increase by 2005 would have to be less than the absolute increase of the past equivalent period, despite all trends upwards. Meeting the Toronto targets seems a long way off and perhaps must rely on fuel-switching, understanding the many limitations of this:

Table 1.3 *Changes by country grouping (Source: PARKER, M.J. (1989)* 'The economics of environmental control of energy'. *Paper given to an* ad hoc *energy forum at Surrey Economics Centre, Guildford, UK. Reproduced by permission of the author and the Centre)*

	Increase, billion tce	Increase, %
OECD	1·58	40
CPE	2·23	130
Other	1·00	135

Table 1.4 *Energy demands and carbon output (Source: PARKER, M.J. (1989)* 'The economics of environmental control of energy'. *Paper given to an* ad hoc *energy forum at Surrey Economics Centre, Guildford, UK. Reproduced by permission of the author and the Centre)*

Energy demand, billion tce	1987	2005
OECD	5·51	7·48
CPEs	3·91	8·35
LDCs	1·74	3·74
Total world	11·16	19·57
World carbon emissions (billion tonnes)	5·78	10·27

(a) Ratios of carbon emission are: coal (at 100), oil 83, natural gas 58, nuclear/hydro 0; coal and oil switching gives little benefit, the two routes left being coal to nuclear, coal to natural gas.

(b) Nuclear is exclusively for electricity and, anyway, coalfired power stations contribute less than 8% of all greenhouse gases; all coalfired generation replaced by nuclear produces modest mitigation of greenhouse effects.

If Toronto targets are to be attempted:

(a) By switching from coal to nuclear, then: (i) world nuclear output needs increasing eight-fold, with about 1500 stations of the size of the UK Sizewell B started by the mid 1990s at the latest; and (ii) world coal output needs reducing by 80%

(b) By switching from coal to natural gas: world natural gas output needs trebling and world coal use completely eliminated, but this is still not sufficient to meet Toronto targets.

Table 1.5 *Recommendations of Toronto Conference (Source: PARKER, M.J. (1989)* 'The economics of environmental control of energy'. *Paper given to an* ad hoc *energy forum at Surrey Economics Centre, Guildford, UK. Reproduced by permission of the author and the Centre)*

	Billion tonnes
Carbon emissions in 1987	5·78
20% target *reduction*	−1·16
Target carbon emissions in 2005	4·62
Projected carbon emissions in 2005, using trends of past 20 years	10·27
Thus total saving required to meet target	5·65
Half each from:	
Energy efficiency/conservation	2·82
Fuel switching	2·82

Table 1.6 *Calorific value of fuels*

Fuel	BTU/pound sterling
Crude oil	18 000–20 000
Gasoline	20 000
Kerosene	20 000
Gas oil (diesel fuel)	19 000
Ethyl alcohol (pure)	11 500
Liquified petroleum gases (LPG)	20 500
Anthracite	13 000
Bituminous steam coal	10 000–15 000
Lignite	6000–7500
Coal briquettes	11 000–14 500
Oil shale	3000–4500
Wood (air dried)	3000–9000
Charcoal	10 000–14 000
Peat (air dried)	6000–9000
Sugar cane bagasse (dry)	8000–9000
Dung (air dried)	4000–5500
Vegetable oil	15 000–17 000
Solar energy	3500 BTU per square meter per maximum

The above hypotheses show the total infeasibility, even globally, of meeting Toronto targets just by fuel-switching, even after large hypothetical savings made by energy efficiency and conservation.

Things look even more intractable after considering global distribution:

(a) Over 80% of world coal is used in ten countries, some with economies dominated by coal production and use; the large coal industries of the USA, the UK and the former USSR republics make vital contributions to their economies.

(b) Two-thirds of world natural gas reserves are in the former USSR republics, plus the Middle East; to transport these to replace coal in the coal-dependent economies is not practical.

(c) The immense increases in nuclear capacity required could not all be absorbed to OECD countries; large numbers would be in developing countries, potentially dangerous for socio-political stability.

1.7 Future energy supply–demand

Information shown in Table 1.6 is at the heart of choice between fuels. Also, something similar to Figure 1.22 must be done for every economy, however difficult or approximate. International continental and national supply-side economic and energy organisations have existed for many years. The demand-side became more highly organised in the 1970s and this will continue. To look ahead it is first instructive to look back. Over the last 100 years, energy showed enormous quantity increases but remarkable price stability, energy per capita increasing steadily over the twentieth century, economic development meaning substituting other forms for human energy. Oil became the major fuel, but it still accounts for only half of total energy, and both its development and production needs substantial capital, with high risks, long lead times and long payback periods. Since all oil producers are making similar decisions continually, large oil volumes become available simultaneously, depressing prices, curtailing investment, restarting the pricing–investment cycle. Five such cycles over the last century seem to be discernible, averaging over 30 years peak-to-trough. Working on this basis, the last investment trough was around 1973; the next can be expected in the mid-1990s, probably influenced by Middle East and Eastern Europe/former USSR conditions. The last oil cycle seemed similar to all previous cycles, i.e. after low investments, supplies tightened, with most marginal oil concentrated in only a few fields, owners attempting to increase economic rents, thus driving up prices. In previous cycles, increased prices quickly led to lower demands and thus relatively modest price movements materialised. However, in the latest cycle demand continued growing despite increasing prices, because of government panic action, changing developing countries' demographic patterns and some strong economic growth persisting temporarily. Large population increases after World War II affected household formation, needing extra housing, transport and energy-intensive goods and services.

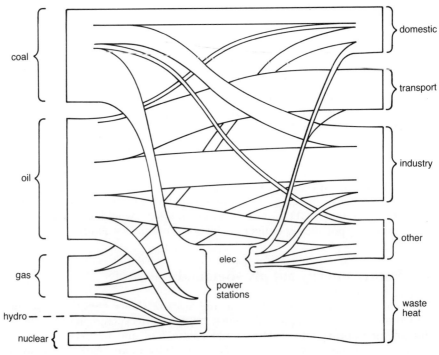

Figure 1.22 *Energy flow in the UK (Source: internal paper, Department of Electrical Engineering, Imperial College (1986). Reproduced by permission of the College)*

Neither demand nor supply positions are similar to the 1970s in the 1990s. In 1973, over 40% of electricity in Europe was from oil; in 1990 this was below 7% because of immense investment in coal and nuclear produced electricity. Large amounts of coal would, in the 1990s, come quickly onto the market in response to increased energy prices and substantial investments have improved fuel flexibility in power stations. On present trends, developing countries' demands will increase by half between 1990 and 2000, by nearly 90% by about 2010, then making up three-fifths of increases in energy, yet still only one third of total energy. Compared with developed countries, developing countries' demand is still too small to have much influence on energy prices until (say) beyond 2020. Yet similar forecasts made for other finite natural resources have proved quite wrong in the past! Conservation forecasters tend to assume long-term energy prices remaining stable, with some fluctuations, but meanwhile the oil cycle described above seems likely to continue, giving rise to over-investment, followed by excess energy and falling prices, indicating modest increases in price from the mid-1990s through 2000 to a downturn about 2005.

1.8 Commentary summary

Countries can be classified into three for development: developed, resource-rich and resource-poor developing. Developing countries have not played a substantial role in total energy on the demand-side but are likely to do so in the medium-term future, partly linked to the increase in private sector funding and influence. World economic outlook is for slower growth (at best $2\frac{1}{2}\%$) post-1990 compared with 3% pre-1990, with many unknowns: developing each country's growth; oil prices and availability; EEC prospects; prospects for the ex-Eastern 'Bloc'; debt servicing; relieving poverty. The likely world energy outlook is backgrounded by these factors and the medium-term dependence of economic growth on energy, especialy in the ex-Eastern 'Bloc' and developing countries. Commercial fuels are abundant, if unevenly distributed throughout the world. Some renewables, e.g. fuelwood are not. All engaged in energy must periodically assess the world economic outlook and all world fuels, nationally indigenous and internationally obtainable. For this, the information needed is wide for each fuel: resource base; exploration rate; rates of extraction and production; economics of production; together with supply risks and restrictions, political, fiscal and operational. Periodically, past and likely future world demands must be assessed in detail for the three major primary fuels in turn, oil, gas, and coal, bearing in mind the role of the MNOCs, energy pricing, interfuel competition, renewables, energy needed for development and dependence of countries on a particular fuel, also, fuel supply and demand balances for the three primary commercial fuels must be assessed, in detail, for the short, medium and long term. Energy management, improved energy efficiency, energy conservation and the effects on the environment will all play an increasing role, but their pursuance will be different in developing countries which have not sufficient resources and in the ex-Eastern 'Bloc'. Especially, energy conservation and environmental matters lack hard economic bases for making proper cost–benefit analysis. Desirable, practical environmental standards will be difficult to achieve worldwide. Future world energy is likely to remain cyclic for price and availability, developing countries not beginning to play a vital role until well after 2000.

1.9 References and further reading

1 BERRIE, T.W. (1983) 'Power system ecnomics' (Peter Peregrinus, London) Chapter 1; also PFEFFERMAN, G.P. and MADARASSY, A. (1989) 'Trends in private investment in thirty developing countries'. Discussion paper no. 6 and updates, International Finance Corporation Economics Department, World Bank
2 International Monetary Fund (1990) 'World economic outlook' pp. 1–4 plus same title (1991) pp. 1–6
3 World Bank (1990) 'World development report, 1990', plus same title for 1991
4 'World Bank News' **IX**, 29, July 19, 1990, plus commentaries in later 'World Bank News'
5 US Bureau of Mines (1980) 'Principles of the mineral resource classification system' plus national UK and USA Geological Surveys
6 ROZA, A.C.A. (1985) 'The management of the oil sector'. Paper presented at the International Symposium on Energy Sector Strategy and Management, Imperial College, London
7 WESLEY, M.W. (1985) 'The management of the gas sector'. Paper presented at the International Symposium on Energy Sector Strategy and Management, Imperial College, London
8 PARKER, M.J. (1985) 'Management of the coal sector'. Paper presented at the International Symposium on Energy Sector Strategy and Management, Imperial College, London
9 MUNASINGHE, M. (1990) 'Energy analysis and policy' (Butterworth Heineman, Oxford)
10 PARKER, M.J. (1989) 'The economics of environmental control of energy'. Paper to Surrey Energy Economics Centre Conference, Surrey University, Guildford

Chapter 2
Electricity sector assessment

2.1 Energy planning

2.1.1 *Background*

Before 1973, with cheap abundant oil, little need existed for energy sector planning, although subsectors—e.g. electricity, gas, coal—often produced local, even national plans. Many countries use unplannable non-commercial energy more than commercial, leading to disastrous fuelwood overconsumption and environmental problems. Assuming, with respect to the 1960s, relatively expensive energy in the 1990s, national energy planning becomes necessary, as no claim in the past that market forces will be sufficient has ever been substantiated. After sudden large oil price rises in the 1970s many countries found balancing energy inputs and outputs was difficult. Even in 1990 [1] serious difficulties arose during the increases in energy sector or subsector efficiencies by pricing, energy management, investment in indigenous energy, to help economies to survive high energy costs, or to build up institutional and management capability to deal with energy sector and subsector shocks [2].

2.1.2 *Improving efficiency*

Most oil-importing countries passed through to consumers both the oil price rises of 1973 and 1979, and no doubt would do so for any future oil price rises. Oil-exporting countries behave more complexly, most pricing oil domestically and in allied markets well below border price. Many developed and developing countries behave like this, reflecting non-border costs and prices into subsectors such as electricity, gas and coal. Although moves existed towards strict international trading or border pricing up to 1980, lowering oil prices in the 1980s virtually stopped this. Worldwide, therefore, there exists room for improving energy efficiency by removing fuel subsidies and imposts, also price imbalances between oil, gas, and coal. This is also true for petroleum-based consumer goods (e.g. plastics), again because indigenous fuels often have socioeconomically set prices much lower than border prices. Improving energy efficiency by energy management or otherwise is described for electricity in Chapters 3 and 4. For large energy consumers (e.g. electricity and transport), government or regulator incentives, plus government or regulator support, financial or otherwise, is needed to encourage savings to the economy, utility and to consumers through improved energy efficiency and energy conservation. Simple improvements are often more effective than capital-intensive schemes for complete, complex equipment replacement. For medium and small consumers, energy conservation measures are often the best way of seeking improved efficiency; i.e. just by using less energy.

2.1.3 *Improving investment*

The total investment required for energy to achieve satisfactory world economic growth is large, even after allowing for improved energy efficiency, energy management and energy conservation. No consistent, comprehensive estimates of energy sector investment needed in developed countries exist, but a working hypothesis suggests between [3] ten and twenty times the needs of developing countries, which is much better known and with a fair degree of consistency. For balanced energy inputs and outputs, and for reasonable economic growth in developing countries, about US$130 billion (1982 dollars) investment over the ten years up to 1993–1995 is needed (these figures may be somewhat out of date but they are consistent and illustrate the points well [1]) doubling energy's share from 2% to 3% of GDP, achieved late-1970s/early-1980s, to an average 5% of GDP over about ten years up to 1993 to 1995. Therefore, no matter how it is viewed, funding global energy is a major problem in the 1990s, as is fitting the best finance, loan, credit, grant, etc., to particular types of energy project within the country and world economic and energy sector circumstances. The scale of developed and developing country energy financing emphasises the need for [4]:

(a) Mobilising every single appropriate financial source, each supplying its own natural share, whether in the private or public sectors
(b) Simplifying making finance available, especially in terms and in agreements on cofinancing packages, presently so complicated that many possible donors and recipients are put off
(c) Strengthening government guarantees, and guarantees by neutral bodies, of financial agreements contractual work, etc.
(d) Providing finance more efficiently, effectively and quickly, for technological developments and technology transfer, particularly for transfers to developing countries.

Energy financing broadly includes the following types:

(a) Official; i.e. from governments, applied multilaterally or bilaterally with respect to countries which are donors and/or recipients
(b) Private; i.e. from commercial banks, finance houses, pension funds, individuals
(c) Suppliers' or buyers' credits; i.e. credits or loans supplied through equipment suppliers, guaranteed or unguaranteed by others, including often governments
(d) Equity; especially for utilities, individuals, organisations or governments taking shares in the corporation or utility.

Investment mixes must fit within the country's national accounts. The US$ 130 billion illustrative figure required for developing country energy quoted above would be much higher without making this latter assumption, especially if most financing was on strict commercial terms. Only small energy investment in developed countries uses foreign exchange, but for developing countries this proportion is large, perhaps up to one half, say by illustration, US$ 65 billion, which must be compared with World Bank figures of US$ 25 billion for the actual flow in 1982 for developing countries' energy. Mobilisation of adequate local developed countries' energy will not prove difficult, but this action again causes problems for developing countries. The favourite method in the past (direct treasury transfers) tends to destroy utilities' autonomy and accountability and distort operations. For this reason and because developing countries' foreign exchange is normally in very short supply, there must be greatly increased, more efficient local capital flows for energy, perhaps only possible if pricing of electricity, gas, coal, and oil reflects true marginal economic resource supply costs and border prices, as mentioned above. But above all else private capital must be welcomed in large amounts into the energy sector to make sure near-adequate capital is forthcoming.

2.2 National economic position

Before dealing with subsectors (e.g. electricity), attention is needed first on the national energy sector and, before this, the world and national economies and the world energy scene (see Chapter 1). It is worthwhile spending time examining the current world and national economy, because of existing integrated international or national economic and energy plans. However, given the tendency in the 1990s towards private utilities' funding and ownership and because of lack of data, a short time only may be possible on the current national economy, but it still must be examined, at least for how best to fit the private sector into national economic and energy policy. Detailed national studies today are a feature of a lot of developed and most developing countries. In many countries outline scenario planning is often considered sufficient. Even governments leaving economic situations mostly to market forces really require (usually through regulators) utilities to piece together *de facto* national economic and energy policies from past government statements and ex-post factor analysis, because utilities must start with at least some idea of how world national economies and energy sectors are being developed and the chances of success of various trends and policies. If there is little published on this, utilities must examine the world and national economic and energy situation for themselves, or employ consultants. Where government ownership and funding for the energy sector and subsector is likely to remain, then some published integrated national economic and energy planning will also remain to be consulted [5].

The World Bank periodically carries out national economic reviews of most developing and some developed countries; many are published, others normally available for bona fide use. Other organisations do similar studies similarly available (e.g. sister organisations to the World Bank, EEC, OECD, USAID, UKODA), and there are reviews done by developed and developing countries themselves for their own and similar usage. All these are useful to peruse for those first embarking on energy sector studies for the first time. For developed countries, OECD and EEC carry out partial and also full national economic studies, perhaps with some commercial confidentiality. It should not normally be unduly difficult, therefore, for anyone working with or for, or taking instructions from, a utility to obtain sufficient background information about how to proceed in this matter.

2.3 National energy and fuels position

2.3.1 Introduction

After looking at the national economy, utilities must look quite closely at the national energy sector, studying any relevant reports available. Many of these reports are written by the same organisations that write national economic reports. If no reports are available, the utility must commission some, giving guidance on what studies are needed.

2.3.2 Energy assessment studies

Any national energy plans that existed in the 1960s and early 1970s were quite complicated and elaborate. However, in the late 1970s there emerged much simpler energy planning in both developed and developing countries, these being favoured by the World Bank who used the simple format in the 1980s to carry out their own energy assessment studies of 70 developing countries. This simple method is recommended for utilities commissioning or carrying out such work in the 1990s [6]. Study topics likely to be generally applicable are:

(a) National demands for commercial and non-commercial fuels
(b) Past, present and likely future international and national supplies of commercial and non-commercial fuels
(c) Past, present and forecasted energy supply and demand balances, internationally, continentally, nationally and locally
(d) Past, present and likely future energy prices, duties, taxes, subsidies, etc.
(e) Energy management, efficiency, conservation, and environmental maintenance policies, existing and likely future, international and national
(f) Energy organisations and institutions which are relevant, international and national
(g) Investment requirements and pricing, in a little detail
(h) Technology transfer factors of particular importance in the circumstances.

2.3.3 Improving energy efficiency

An important follow-up activity to any energy studies is studying how to improve energy production, transmission and usage efficiency, in every way possible which is economic, because these give very high returns on the capital and labour allocated for this purpose. The World Bank format for listing improvements in energy efficiency (the term being used in the widest sense) can be copied with advantage by most, and is as follows:

(a) Establish approximate national energy efficiency capabilities including: manpower development, organisational strength, institutions, energy management activity, and study energy audits for industrial, commercial and residential consumers.
(b) Prepare preliminary studies of potential energy-saving investments in all subsectors, especially where savings are expected from other energy

sectors' or subsectors' experience, to pinpoint likely energy conservation projects of the future.
(c) Identify particularly any means for energy loss reduction, e.g. by improvement of the efficiency of electricity generation, transmission and distribution.
(d) Identify renewable energy sources plus the institutions needed to exploit them, also suitable policies and practical programmes to develop such renewables.
(e) Establish where shortages exist of adequately trained energy policy-makers, managers, analysts, planners and technicians which, together with organisational and managerial shortcomings, act as major constraints on having efficient and effective energy sectors.

2.4 Ownership patterns

Up to World War II, traditional utility ownership patterns globally were a combination of private, municipal and local power boards, with foreign franchises in some developed and many developing countries. Except mainly for North America, this ownership pattern was replaced by national public ownership and funding, this only starting to be changed again worldwide in the late 1980s. Changes in ownership profoundly affect planning and pricing [7]. The 1990s will see a steady change from publicly to privately owned and funded utilities, with competition between producers, especially in electricity which will be increasingly treated as a commodity, with the 'spot' pricing, buying forward, future markets, agents, brokers, etc., which goes with this [8]. The major effect to be noted here of the change from public to private funding and ownership is the change from optimising investment and pricing in accordance with what is best for the national economy to what is best for the utility, and from long-term economic to short-term financial optimisation criteria. All these effects are examined in later chapters.

2.5 National electricity position

Following the carrying out of national energy reviews, as described above, utilities need to make a similar study of the national electricity position. With the trend towards privatisation mentioned above, because one utility will not be the norm for the 1990s, national electricity studies follow similar approaches to national energy reviews. Electricity is a most important factor in each and every economy for promoting all development. In 1990 total electricity consumption in developing countries was one-twenty-fifth of that in developed countries, giving a gap twice as large as the gap between these types of country in GDP *per capita*. However, with enormous developing country funds allocated to electricity, and with imminent saturation of developed country electricity demand due to demand management, efficiency and conservation, these gaps will decrease in the 1990s. There are two national electricity dimensions to consider in any study—supply and demand—although there are often difficulties. With supply, studies deal with designing, financing, constructing, operating and regulating the electricity subsector, bringing in organisations other than those making up the subsector, because relationships between electricity and overall energy supply may often be in turmoil in the 1990s due to privatisation. On the demand side, difficulties will exist because consumers of all types—residential, commercial, agricultural, industrial, etc.—will be in the electricity market also. Again, many electricity markets may be in turmoil due to privatisation.

In most developing countries, efficient electricity markets are unlikely in the 1990s, and development programmes must needs compromise between traditional long-term economic and modern short-term financial optima, many factors applying: likely funding; urban versus rural development priorities; electricity service quality expected; technology transfers, etc. Socio-economic factors will sometimes be paramount, e.g. rural electrification, life-line tariffs, and environmental maintenance. In developed countries, and beyond 2005 in many developing ones, national electricity markets will increasingly be determined by demand management, to the theoretical point where the level of total electricity demand is what suppliers and consumers jointly want it to be, at any instant, using interactive load control [9]; Chapters 3 and 6 develop this further. However, a number of supply utilities will always compete for market shares, making it difficult for them to dictate market levels, or clear markets jointly either with each other or with consumers, let alone come to an agreement to clear electricity markets to the mutual advantages of suppliers and all consumers. Nevertheless, each supply utility must decide what market share it is aiming for, and at what prices (as must each retail utility), from its own point of view. Much more research is needed into how to install efficiently interactive load control for all suppliers, retailers and consumers.

Single national utilities will find it worthwhile to fully study the national electricity markets, wholesale and retail. Separate utilities will do more studies, but of the same type. Whoever is doing national electricity studies it is important to:

(a) Identify likely principle national problems and constraints for the short, medium, and long term
(b) Analyse past and existing goals for national electricity, determining likely future goal scenarios, bearing in mind existing and likely problems and constraints, identified above
(c) Determine scenarios of likely changes in institutions, policies, practices, to achieve the scenarios or goals determined above.

Normally further studies are needed [10] before actual development programmes of individually named projects are isolated out from the national electricity background, ranging through preliminary engineering and feasibility studies to studies for structure and management of utilities, financing plans, financial sources, effects on utility accounts, legal questions, pricing, economics, technology transfer and training. Some of these themes are further developed in Chapter 7. National electricity studies provide excellent background data for all utilities, providing likely practical scenarios, indicating meaningful, alternative approaches for utilities and government, national planners, shareholders, financiers, regulators, consumers, and consumer consultative councils, etc. They also help modellers.

2.6 Electricity markets position

2.6.1 Introduction

Worldwide, privatisation is claimed to improve markets. Not much electricity markets efficiency testing was done until the 1980s, electricity being so often regarded as a public service, a public good, and natural monopoly which, therefore [11], should be publicly owned and/or heavily regulated. Welfare economists developed most existing methodologies for electricity investment, planning and pricing, believing that, to maximise electricity markets efficiently, investment must be in accordance with optimised, least-cost, long-term economic development programmes, and prices then set in accordance with the long-run marginal costs (LMC) of that programme [12]. The challenge to this came in the 1980s when the desirability of holding electricity in public hands became seriously challenged, coinciding with attempts to get private finance into what had, up to then, been considered the public domain.

2.6.2 Markets

The welfare economist's approach claimed to produce efficient wholesale markets [13]. Circumstances meant virtually no electricity producers' competition from auto-generators, etc., because of price 'penalties' charged by those producers for such supplies not being directly controllable by the latter for availability at peak. With 'clever' meters, microcomputers, etc., and time-of-use, dynamic and real-time 'spot' pricing (Chapter 6), in the 1990s wholesale marketplace efficiency can be improved using short-run marginal financial costs (SMC) tariffs, set near the time of actual use rather than LMC economic tariffs prescribed well in advance of usage, from long-term economic development programmes, as at present. Wholesale market efficiency could then be increased by competition between producers, autogenerators and cogenerators, the latter being offered SMC buy-back prices whenever they sell electricity. Eventually, wholesale electricity will be treatable as a commodity.

Welfare economics found retail markets difficult to prescribe for because retail LMC are not physically meaningful. Most equipment needed for retail SMC, time-of-use, etc., pricing is already developed and trials done. Larger-scale usage will bring economies of scale and cost decreases in the 1990s. Although rarely can there be efficient retail electricity competition, using the tariffs described above must mean more efficient retail markets [14]. This subject is taken up again in Chapters 5 and 6.

2.6.3 Market assessments

Electricity sector assessments, whoever does them, must evaluate existing and likely future status, scope, competitiveness, efficiency, etc., of wholesale and retail markets, asking such questions as:

(a) What are the market pricing principles, LMC, SMC, prescribed, time-of-use, dynamic 'spot' pricing? What should these objectives be?
(b) How much competition exists from autogenerators, cogenerators, or between producer utilities?
(c) What regulatory authority(ies) exist(s); do these affect market efficiency?

44 *Electricity sector assessment*

(d) What are the patterns, scope, range, etc., of wholesale and retail prices; are the two related?
(e) What gaps exist in data needed to do marketplace assessments, especially for developing countries?

2.6.4 *Consumer reaction*

Wholesale and retail consumers worldwide react, sometimes vigorously, to tariff changes; perspective in this matter can be obtained by asking before making changes:

(a) How often have tariffs changed and with any pattern in the past?
(b) How did wholesale or retail customers then react short term and long term?
(c) Did utilities learn from such consumers' reaction, short term and long term?
(d) What type(s) of tariffs changes caused the most significant reaction?
(e) Did utilities' ownership changes alter tariffs or consumers' reaction?

It is important to assess periodically whether consumers, directly or as seen through their behaviour, are conscious of market efficiency. In theory, the more efficient the electricity markets—i.e. the more competition there exists, which consumers can fully respond to—the nearer is 'pareto optimality' approached, i.e. maximum joint supply and demand net economic benefit. Although it is only wholesale markets which benefit in practice from direct competition, retail markets can improve by:

(a) Notifying and displaying more prominently time-of-use, dynamic or 'spot' prices to retail consumers so that they or their simple automatic microprocessors, can decide on whether to use electricity for a particular purpose at a particular time
(b) Keeping this notification clear, simple and omnipresent
(c) Educating consumers to understand new tariffs
(d) Finding ways for retail consumers to purchase or rent 'clever' meters, microcontrollers, etc., to fully use SMC, time-of-use and dynamic pricing.

2.6.5 *Competition from other fuels*

Parts of 'pareto optimality' above require near perfect competition for all energy subsectors, not possible because of: (i) inherent physical differences between fuels; (ii) inability of consumers—especially retail—quickly or financially efficiently to change their apparatus; and (iii) national fuel subsidies, quotas, etc. However, any national electricity assessment must recognise competition from other fuels, asking the questions:

(a) How does government policy on fuel subsidies, quotas, preferences, availability, etc., affect competitiveness of electricity?
(b) How substitutable are other fuels with electricity, price-wise, physically, socially, environmentally?
(c) How are consumer reactions different for price, availability, quality, etc., for changes in fuels?
(d) What are the effects of MNOCs and other fuel agencies, directly or indirectly, on fuel competitiveness?

(e)   What are the effects of ownership changes directly or indirectly, on competitiveness?

2.7 Individual utility position

Each utility not having access to current assessments of national economy, national energy and national electricity, must make some of their own studies in the manner described in the appropriate sections above, before making detailed assessment of their own individual utility. Having done this the utility must ask what are their own utility's:

(a)   Market share of electricity, wholesale and/or retail?
(b)   Composition of market shares between types of production and/or classes of consumer, by time of day, etc.?
(c)   Price competitiveness, *vis-a-vis* other fuels and other electricity utilities, by tariff types and consumer classes?
(d)   Ownership and funding pattern, compared to other utilities?
(e)   Financial viability, e.g. annual return on assets, self-financing ratios, etc., compared with other utilities?
(f)   Efficiency, managerially, institutionally and operationally, compared with other utilities, measured by; staffing ratios, output ratios, accounts receivable, etc.

Each utility must consider scenarios from which to choose its policies concerning fuels, planning, investment, pricing, market penetration, etc., realising policies might need changing at short notice. Each scenario needs spelling out so as to be easily referred to, quickly modified and easily displayed, recorded and described to all concerned.

2.8 Pricing assessments

Pricing issues are an important part of electricity sector assessment. Carrying out pricing assessment means:

(a)   Studying past and likely future types of pricing prevalent nationally and in the utility in question
(b)   Assessing past and likely future pricing in other utilities, rivals or otherwise
(c)   Carrying out surveys of the utility's consumers by type and class, determining alternative scenarios of future utility pricing and also of the pricing in other utilities especially if in competition
(d)   Analysing utility's consumers' reactions to previous tariff changes, by consumer type and class.

Within any pricing policy there are many ways of checking on whether the revenue recovery will provide adequate cash flow (see Chapter 6).

2.9 Future market shares

All electricity utilities competing in the wholesale markets by feeding into national or regional grid must assess both the size of the total market and what market shares will be at present and in the future, remembering likely world and national economic growth, world and national energy growth, improving energy management, efficiency, conservation, and better environmental maintenance. The end result will be likely future scenarios of total electricity and market shares. Estimating retail market shares is simple compared with wholesale shares because of lack of retail competition and the few practical arrangements possible at the retail level, needing only some method for forecasting deviations from past trends (Chapter 3), e.g. due to changes in ownership, funding, investment and pricing policy.

With growing demand management, theoretically to where markets are always fully controllable (eventually jointly by suppliers and consumers to the mutual benefit of both), it becomes possible for upper and lower limits to be set upon instantaneous demand, and for the market always to be cleared efficiently between these levels. This gives new dimensions to pricing, investments, planning, operation, competition, market shares, and the market dimensions become internalised to pricing and planning. Meanwhile, each market is different and finding optimum market shares for a particular time, within a variable total market, will always be complex. Much learning will be by experience only, rival producers determining their optimum market shares by exercising demand management in their own way with wholesale consumers. There will also be a learning process for retail utilities, exercising demand management with retail customers and to various extents, theoretically downwards to zero loading and upwards to full utilisation of total consumer connected apparatus at any point in time for all, or for a particular type of consumer, at market levels which clear most efficiently, as found by experience.

2.10 Regulators

Before the 1980s, regulation of most utilities was automatically carried out by government, most utilities being owned and funded by governments, who were responsible for policy, planning and pricing but not day-to-day operation. The exceptions were mainly in North America, where federal, state and local regulators operate in what is often described as 'heavy' regulation because of the regulator's vast powers. With emphasis on private funding and ownership more regulatory bodies will be set up by governments, but operating independently, these varying in character. This theme is pursued in more detail in Chapter 5. Electricity sector assessments must pursue existing and likely future regulatory positions, starting with their duties, which are:

(a) Pricing; ensuring equitable, comprehensive, fair, meaningful tariffs
(b) Service quality; ensuring adequate consumer service, including reasonable standards of supply
(c) Financial; ensuring normal commercial accounting standards

(d) Business; ensuring normal rules on advertising, hiring and firing, local social obligations, etc.
(e) Externalities; ensuring environmental maintenance, conservation, efficiency measures.

For these functions, regulators have powers whose scope and application depend on whether regulation is 'heavy' or 'light'.

Light regulation means:

(a) Regulating unobtrusively, when and only to the extent that is really necessary, keeping in the background
(b) Relying on normal internal utility self-regulation for day-to-day duties, and/or on market forces
(c) Regulating only as a last resort in blatant cases of abuse
(d) Having a small staff, hiring in specialists when required.

Heavy regulation means:

(a) Having a prominent appearance
(b) Expecting frequent, early consultation by utilities for price rises, or anything else considered important by the regulator
(c) By-passing the normal machinery of utility decision-making for important reasons
(d) Sending regularly for people and papers, development programmes, investment decisions, financing plans, etc.
(e) Having many specialists and a large staff with high workloads.

Regulatory bodies are normally set up and appointed by government, but should operate independently, those in charge being recruited from a cross-section of expertise and experience (engineers, economists, accountants, business experts, civil servants and academics). At least some should have had experience in electricity and energy. When carrying out electricity sector assessments the following questions on regulation are relevant:

(a) Which regulating body(ies) is concerned with electricity? Which should be?
(b) What type of body is this, direct government, heavy or light regulation? How many staff has it? How many specialists? What is its impact?
(c) What are the terms of reference for the regulator? What matters does it deal with, e.g. pricing, investment, financing plans, etc.?
(d) What else is known about the regulator, e.g. past record of dealing with utilities, degrees or investigations in depth?
(e) Is the degree of existing regulation likely to become more or less severe, extended or contracted? Is this justified from past experience?
(f) If no regulatory body exists, should one be set up, what type, with what duties?

2.11 Commentary summary

National electricity assessments must be done by all utilities from time to time to retain perspective. In sectors made up of one, probably government funded and owned, utility, then national electricity surveys become part of national economic and national energy surveys. Even with plural utilities whether wholesale or retail, some national electricity assessment is necessary, the benefits always exceeding the costs, although the amount of work worthwhile will vary with the number, type and importance of the utilities. Electricity reviews must start with assessments of the national economy and national energy, to a depth sufficient to enable separate electricity utilities to make comprehensive, meaningful decisions concerning their own investment, planning, pricing, operation, financing, revenue requirements, returns on assets, etc., *vis-a-vis* the rest of the economy. Privately owned utilities may well find it difficult to obtain information about the national economy and national energy, and probably will hire experts to do this work. A careful look must be made at the number and kinds of utilities, producers, transmitters and distributors, their ownership and funding, with any present trends in these, how quickly these trends are changing, and what this means with respect to national electricity. Such reviews enable utilities to make separately detailed, meaningful, assessments, of national electricity, past and future, its growth rate, funding organisation, format, etc.

Utilities must then look at current and likely future consumers' position and the electricity marketplaces, asking such questions as: how efficient are wholesale and retail electricity markets, what competition exists, how are electricity prices set, who regulates markets and how well? The question of consumers' reaction, past and likely future, to changes in price and electricity quality is always highly relevant. Each and every utility must then carefully examine its own position in national electricity, its share of the wholesale and/or retail markets, and prospects for the future of: total electricity markets growing or shrinking or changing in nature; also, utility share of markets' growth or shrinkage or changing. An examination is needed of utility fuels' position; just how secure are supplies of primary fuels, and their availability and price, past, present and future?

Current and future electricity pricing warrants a utility survey of how far modern pricing is used and likely to be made, e.g. 'clever' meters, microprocessors, microcontrollers, microcomputers, enabling demand management, time-of-use tariffs, dynamic and 'spot' pricing to be introduced, wholesale and possibly retail? The whole future of electricity pricing *vis-a-vis* other fuels needs considering in the study, also pricing and load management of policies of individual utilities, especially rivals. Indeed, the 1990s electricity market place(s) position becomes vital to all assessments, given competition, at least in production, private funding and ownership. When demand management coupled with dynamic pricing becomes normal, demand can theoretically be completely controlled by physical means and pricing and electricity markets cleared efficiently at every instant in time, at a level of demand eventually jointly beneficial to producers and users. Market shares are then no more important than the total market size.

The influence, jurisdiction and scope of regulatory bodies, presently and in the future, must be assessed in any electricity review. Regulatory bodies will vary between full direct government control over ownership of and supplying funding for utilities, through independent, but government appointed heavy regulatory bodies with authority to send for persons and papers, to independent light regulatory bodies which keep a low profile, only intervening for flagrant abuses in pricing or markets efficiency, to ensure maximum competition, relying on utility self-regulation as much as possible. Regulatory bodies' jurisdiction and scope will be important in the future to all utilities, which must take them into account when investing, planning and pricing, gauging whether they will intervene in some areas, e.g. when using the grid as a common carrier, ensuring autogenerators and cogenerators are given full scope, whilst still getting fair prices for their electricity.

2.12 References and further reading

1 WORLD BANK (1983) 'Energy transition in developing countries', and later publications on the same theme by the Industry and Energy Department of the World Bank, Washington DC, USA
2 WORLD BANK/United Nations Development Programme (UNDP) (1983) 'Energy sector assessment programme' quarterly reports since 1983
3 SPENCER, R.T. 'World energy needs', *Energy International Journal* **VIII**, 1982, p. 32
4 WORLD BANK (1983 and revisions) 'Cofinancing'
5 MUNASINGHE, M. (1980) 'Energy analysis and policy', also 'Electric power economics' (Butterworth Heineman, Oxford); also BERRIE, T.W. (1983) 'Power system economics' (Peter Peregrinus Ltd., London)
6 Quarterly reports of the World Bank/UNDP Joint Energy Sector Assessment Programme published since 1980, from the Energy and Industry Department, The World Bank, Washington DC USA
7 BERRIE, T.W. 'The new power sector planning', *Energy Policy*, October 1988, pp. 453–457; also BERRIE, T.W. and McGLADE, D. 'Electricity planning in the 1990's', *Utility Policy*, **1**(3), April 1991, pp. 199–211; also the 1989 published 'Report of the Task Force' commissioned by the New Zealand Government to look into the organisation of the N.Z. electricity supply industry, New Zealand House, Haymarket, London
8 BERRIE, T.W. and HOYLE, M. (1985) 'Treating energy as a commodity', *Energy Policy*, Dec. 1985, p. 506
9 BERRIE, T.W. 'Interactive load control', series in *Electrical Review*, September 1981 to June 1982
10 BERRIE, T.W. (1983) 'Power system economics', Chapters 2, 5, 8, 9, 10 (Peter Peregrinus Ltd., London)
11 ROTH, G. (1987) 'The private provision of public services in developing countries'. Published for the World Bank, Oxford University Press; also BERRIE, T.W. 'Privatisation in developing countries'. Crown Agents Review 1989, Vol. 1
12 TURVEY, R. and ANDERSON, D. (1977) 'Electricity economics'. A World Bank research publication, the Johns Hopkins University Press
13 Central Electricity Generating Board (1986) 'History of the bulk supply tariff'
14 SCHWEPPE, F. *et al.* (1988) 'Spot pricing of electricity' (Kluwer Press, Cambridge Massachussetts USA)

Chapter 3
Demand forecasting, management and reliability

3A DEMAND FORECASTING

3A.1 Traditional forecasting

3A.1.1 Introduction

Traditionally [1] forecasting is at the heart of electricity planning [2]. Given the long plant lead times and economic lives in electricity in the immediate past, forecasts were needed for five to twenty-five years ahead, of both peak demand (kW) and patterns of demand (Figure 3.1) either:

(a) forecasting peak demand (kW), then using forecasted load factors deducing annual kWh; or
(b) forecasting annual kWh by consumer class and summating, then using forecasted load factors to deduce kW; or
(c) forecasting kW and annual kWh separately, checking derived load factors against experience; or
(d) forecasting annual national economic and energy sector growth, then breaking down the latter into subsectors deducing annual kWh, then converting kWh into kW using forecasted load factors.

Traditional forecasting methodologies are:

(a) Extrapolation; e.g. continuing previous growth until saturation is allowed for, examining time series of kW and kWh, then reviewing fuel and plant prices, likely new industries, improved energy efficiency, demand management, conservation and environmental measures, to be able to modify the trend sensibly
(b) Synthesising, e.g. factors behind demand are examined, predicting annual kWh and demand patterns for future economic scenarios, using correlations of electricity growth with population, per capita income, consumer numbers, type and class, appliances sales, industrial production, etc. Table 3.1 illustrates.
(c) Market surveys; e.g. kWh and demand patterns examined using various credible tariffs, for existing and new residential communities, industrial or commercial estates, etc.

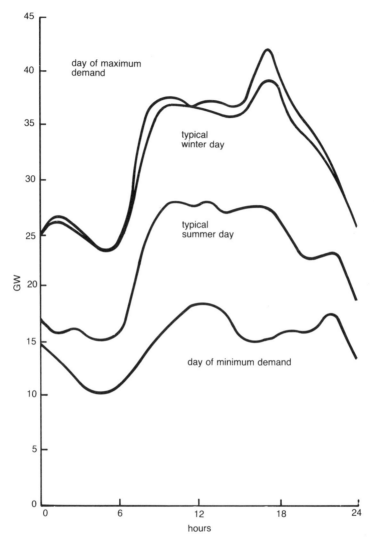

Figure 3.1 *Typical daily demand curves*

3A.1.2 National economy and energy

Before forecasting electricity growth, world, national economic and energy growths are examined, using interlinking models (Figure 3.2), allowing for elasticities. However, past elaborate predictions have often proved to be unsatisfactory [3] because they assumed unsustainable trends, did not detect trend changes (e.g. predicting US$100 per barrel oil), ignored consumer reactions to price, and neglected technological breakthroughs and new energy sources. UK energy forecasting provides examples [4], in 1966 predicting rapid coal depletion, deterministic planning in its heyday being preoccupied with optimal fuel mixes to meet forecasted energy demand. Assuming cheap, abundant oil and electricity production using coal, oil, gas and nuclear, forecasting errors were considered compensatable just by varying imports of cheap oil. Nuclear could be expanded rapidly, reliably and competitively with oil, again implying additional cheap energy. Such forecasts were aggregate, using 'top-down' approaches only, energy estimated as a function of GDP in primary, not in end-use, terms. Forecasts for a short time being frequently correct, unequivocal assumptions about exponential energy growth became commonplace in many countries, e.g. the concept of demand doubling every seven to ten years. Specialists became increasingly uneasy in the 1970s about aggregated, top downwards, exponential energy forecasting, not allowing for much saturation and it is now accepted that elasticities between growth in GDP and economic growth alone should not be used for forecasting, especially for competitive fuels such as electricity.

The 1970s and 1980s witnessed rapid evolution in international and national energy forecasting:

(a) Using longer time horizons, 20 to 50 years; annual forecasts checked for consistency

Table 3.1 *Crude elasticities of electricity demand in a sample of countries: 1970–1975 (Sources: 1977 World Bank Atlas and UN Energy Supply Statistics, Series J, 1973 and 1975)*

	GNP per capita	Annual growth rate of GNP 1970–75	Annual growth rate of electricity consumption 1970–75	Elasticity of electricity consumption
	(US$)	%	%	
India	140	2·6	7·1	2·7
Sweden	8150	2·7	5·8	2·2
USA	7120	2·4	4·1	1·7
Canada	6930	4·7	5·9	1·3
Norway	6760	4·0	5·8	1·5
Federal Republic of Germany	6670	2·1	4·4	2·1
France	5950	4·2	5·6	1·3
Japan	4450	5·3	6·3	1·2
United Kingdom	3780	2·2	1·8	0·8
Yugoslavia	1550	6·8	9·0	1·3
Brazil	1030	9·1	11·6	1·3
Taiwan	930	7·7	10·4	1·3
Thailand	350	6·5	14·6	2·2
Kenya	220	5·9	14·9	2·5
Indonesia	220	5·9	11·7	2·0
Sri Lanka	190	2·8	6·6	2·3
Pakistan	160	3·8	5·3	1·4
Afghanistan	150	4·3	11·8	2·8

Demand forecasting, management and reliability 53

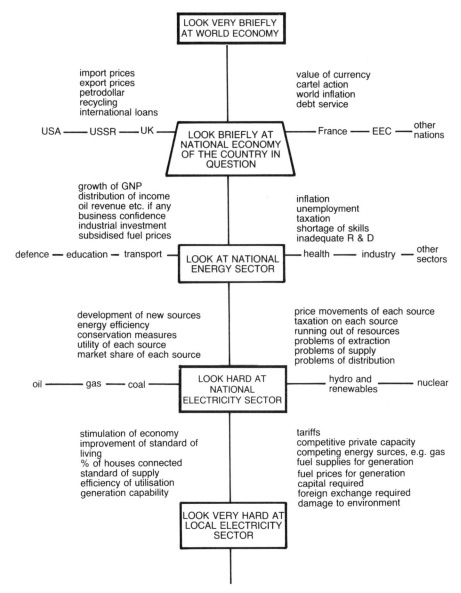

Figure 3.2 *Factors affecting electricity demand*

(b) Models assimilating large databases, permitting rapid sensitivity analysis and alternative scenarios examination
(c) Measuring energy in end-usage terms
(d) Using also 'bottom up' approaches, avoiding elasticities, allowing for fuel competition.

Fuel substitution matters considerably in electricity forecasting, underlying influences being relative fuel prices and availabilities. Technology transfer provides insights into lead times necessary and practicalities of fuel substitution. Substitution of oil is common, then gas for coal, or vice versa, in electricity and in industry. Originally it took 100 years for oil to replace coal as the world fuel. Future substitutions will occur more rapidly, especially where supply infrastructures exist, e.g. gas conversions. Nevertheless, timescale will be important, especially for establishing an all-electric future, given slower world economic growth compared with the immediate past, also greater market uncertainties, high real interest rates and lower capital availability, especially for publicly owned utilities.

3A.1.3 National electricity

National electricity forecasts are approachable by 'top-down' disaggregation, using mixtures of the traditional methods already described [5]:

(a) Disaggregating energy consumption into subsectors (electricity, etc.) within each economic sector; formulating both time-series and cross-sectional data; statistically analysing the past to make forecasts, also forecasting from cross-sections using 'bottom-up' approaches; building national electricity sales from disaggregated statistics

(b) Selecting for each consumer type or class 'activity indicators'; variables expressing sector activity levels which influence electricity consumption, e.g. industrial output affecting electricity for motive power and process heat; amount of office floor space affecting heat; consumers' real incomes; or influencing ownership of electrical appliances usage; or assessing historical relationships between each activity indicator and electricity consumption in each economic sector, then forecasting each indicator, then electricity consumption, assuming efficiency improvements

(c) Selecting competitive-position factors *vis-a-vis* electricity and other fuels (e.g. relative prices and technical characteristics); assessing the effects of these

(d) Applying electricity market share forecasts in each economic sector to total energy forecast in that sector; estimating total sector electricity sales, leading to utility market shares analyses (below).

Traditional electricity forecasting methods grew up for publicly owned utilities and their applicability after utility privatisations worldwide is unknown. It would seem that for the 1990s:

(a) More reliance will be needed on judgement, forecasts made under public ownership being misleading after privatisation because changed incentives and emphases will result in different growth rates, usage patterns, etc.

(b) Publicly owned utilities had 'cost-plus' environments for projecting demand, building plant to always satisfy that demand, passing on all costs to consumers, even when assuming these to be 'least-cost'; privately owned utilities have no cost-plus atmosphere

(c) Privatisation will affect total sales and maximum demands, not merely proportions met by different producers, utilities no longer building up capacity necessary to meet predictable share of demand, pricing and other marketing instruments; this becoming paramount with autogeneration and cogeneration present.

3A.1.4 *Market shares*

Individual wholesale and retail utilities, cogenerators and autogenerators within national electricity sectors must forecast market shares, starting with market forecasting, a traditional electricity commercial approach using experts who know the sector well and which has changed little over 50 years, remaining attractive in the 1990s because it:

(*a*) Resembles methods used in industry and is understandable to entrepreneurs, businessmen and consumers alike
(*b*) Uses salesmen knowing the sector and market well
(*c*) Allows 'bottom up' approaches to be added, trending loads of different consumer types, correlating electricity with incomes, etc., doing market surveys independently by different, interdependent utilities, but with some background commonality needed from government or regulators if any are to be meaningful
(*d*) Ensures usage of macro-economic assessments about national and sector growth and national energy policy; vital to forecasting by private utilities.

However, improvements are needed for the 1990s. First, threshold per capita income must be known for assessing minimal electricity purchases, an integral part of pricing but also forecasting in poorer societies. Second, also needed are not only consumer preferences, which are addressed in market surveys, but also various fuels' demand against price elasticities and cross-elasticities, especially for lower income and new consumers. Such knowledge will allow time horizons in market forecasting to be extended beyond the present short period of one to five years.

Basing market forecasts upon short-time financial data for fuels rather than long-term economic data leads to distortions with respect to the national economy, possibly to utilities:

(*a*) Autoproduction often looks attractive financially but may be nationally unattractive using long-term economic resource costs.
(*b*) Substituting other fuels by electricity often looks attractive financially but may not make sense on a resource-costs basis.
(*c*) Electricity drives often look attractive financially, but possibly not replacing other drives if resource costs are considered.
(*d*) In rural areas, substituting according to financial rules electricity for kerosene lighting, heating, etc., may not be attractive at all using resource costs.

To make sound market forecasts and inspire confidence in government and investors, especially in developing countries, it is necessary to present the long-term indications as well as the short-term and to integrate credibly established and any 'green-field' supply areas, electricity requirements for the latter often being much less than for the former in the short but not the long-term, especially in poorer countries and regions. When old and new areas share the same generation, transmission and distribution, their load forecasts can be integrated with more credibility than otherwise is the case.

The other type of market share forecasting is synthesising from econometric data as touched upon earlier. Such econometric forecasting depends heavily upon valid input–output tables for the whole of the economy and knowing certain, not always reliable, economic ratios; in any case further improvement in the following practices is needed for econometric forecasting:

(a) Data collection, analysis and sorting
(b) Econometric modelling, however simple, of credible disaggregations, from the macro-economic (i.e. beyond the energy level) down to subsectors gas, coal, oil, electricity, etc.
(c) Multi-sectoral models operating more easily and with more stability, better interaction of macro-economic, energy and electricity appliance usage models.

3A.1.5 *Commentary summary*

Traditional electricity planning starts with demand forecasting using semi-exogenous data on cost and price, finding the least-cost way of supplying the forecasted demand, costing this, then pricing those electricity supplies accordingly; deciding whether consumers will pay the long-term price; wondering whether the utility can find the capital or collect sufficient annual revenue, and finally, when the answer to any of these questions is negative, repeating the cycle as many times as necessary to get all answers positive. Bad load forecasts are economically detrimental because of vast resource costs losses and because pricing changes are socio-politically difficult, especially if sudden and upwards. Principles behind forecasting are researching historical statistics and social, economic and political pointers showing how the economy, especially the energy and electricity sectors, are developing. Data sources are development agencies', economic communities' and individual government's economic reports, energy reports, electricity sector reviews and interviews with people in the country, community, and sector concerned. Derived factors such as electricity growth, sales per consumer, etc., must be compared with statistics from other countries and sectors; uncertainties must be acknowledged and cross-checking done by making more than one type of forecast. Common pitfalls are: extrapolation of trends regardless of distorting factors existing; forecasts influenced by supply-side domination (i.e. preconceived ideas of size and composition of generation programmes needed and competition); not questioning whether the economy or utility has capital, skilled labour or other resource costs to meet the forecast; or whether consumers are willing and/or able to pay the price; also excessive 'number crunching' for its own sake.

The three traditional methodologies for forecasting are to some extent interrelated: (a) extrapolation of both historical maximum demand (kW) and annual kWh, bearing in mind trend distorters (e.g. saturation); (b) synthesising annual kWh consumed from econometric data and using forecast load factors to estimate kW; and (c) market forecasts by consumer class and type, again using forecast load factors to obtain kW.

3B DEMAND MANAGEMENT AND RELIABILITY

3B.1 Introduction

Demand management (DM), deliberately influencing demand to optimise [6] supply, in its oldest form, cuts off electricity at a prescribed demand level. For the 1990s DM is vital, DM often being cheaper than providing more electricity supply. Historically, supply shortages meant direct action on selected consumers, quickly remedied each time by installing high energy-cost peaking plant or persuading consumers to defer demand. DM also improved and regulated forecasting ills, encouraging equitable thermal generator loading, allowing constant generator output over long periods at maximum efficiency.

Optimum reliability occurs when marginal benefits of incrementally increasing reliability are less than marginal costs, traditionally dealt with when planning least-cost generation and network programmes [7]. DM reduces reliability for those just joining but forces consumers to give priorities to loads, assigning lower priorities to DM loads, recognising short- and medium-term fuel switching. Important always is consumer response to any reliability changes just as much as price changes, found by consumer load research. Because DM applies to all consumer types and classes for the 1990s, getting optimum response requires consumer education plus incentives, pricing or otherwise (e.g. improved reliability). Utilities, increasingly using short-term financial rather than the traditional long-term economic criteria, will present consumers with alternative prices and other benefits for differing DM schemes, consumers assessing their own reliability and welfare requirements and aspirations.

3B.2 DM by apparatus

3B.2.1 Demand curtailment

Directly switching off consumer 'blocks' at high voltage is effective, but with little discrimination. Yet it prevents system collapse following major plant loss. Selective consumer-contracted load shedding remains satisfactory for limited control over peak demand. Such 'interruptible' load DM is today common, consumers reducing to a predetermined demand at utilities' request (requiring reliable communication); this is always justifiable for large consumers, who for an incentive can stand interruptions for an hour. Again for large consumers, peak tariffs providing DM back-up incentives are possible; in the 1990s devices are readily available for successfully anticipating times of peak and accordingly adjusting consumer demand for particular industrial processes, artificial intelligent systems learning from the past, getting better and better at adapting demand to times of cheaper electricity supply.

Consumption is then such that traditional load shedding is never imposed. To be used similarly in industrial processes is commercial heating which has storage in buildings and boilers, maximum benefit coming from hot storage lasting about a day. With air conditioning, 10 to 20 minute off-periods go unnoticed and controlled off–on cycling makes DM worthwhile. Ventilating fans are also similarly cyclicable, provided occupants are unaware of background changes. Commercial lighting DM mainly depends upon daylight levels and switching off lights. Up to one hour per day decreased lighting is permissible in architectural displays, building illuminations, etc., controlling these at peak being worthwhile.

Domestic consumers have DM precursors in existing time clocks with one off-peak period every day. Modern DM directly controls sheddable domestic loads, treating each consumer separately for water heaters, storage space heaters and freezers. Simple, cheap thermal ceramic storage with time constants between 12 and 17 hours provides controllable loads. Water heating employs large thermally lagged tanks, storing enough hot water during the 'off' time. Freezers and refrigerators are also controllable domestic loads, particularly with additional thermal insulation and larger compressors. Cooling media storage for air conditioners also helps DM.

3B.2.2 DM effects

Figure 3.3 shows five DM effects. Singled out for mention here because of its importance, peak clipping is widespread for reasons already given; 10% peak reduction is realistic against developing country load growth of 5% to 15% p.a., but two aspects are important; 'restrike demand' (i.e. load increasing after curtailment to values higher than the clipped peak because some devices require replenishment after DM, this encouraging DM staggering); also non-recoverable loads (because overall energy consumption reductions due to DM can be difficult if DM brings revenues below financial targets, revised revenue forecasts assuming DM then being required). Large consumers might switch to alternative fuels following DM or increase conservation, or improve processing efficiency, reducing peak plus total demand for electricity.

3B.2.3 DM communication

Three communications categories exist for direct DM:

(1) Radio teleswitches
(2) Telephone or pilot wire
(3) Mainsborne

It is advisable for utilities to carry out trials before starting comprehensive DM [8] because one utility's experience needs careful study before application elsewhere. In radio teleswitching, central stations broadcast coded signals over wide areas using integrated circuits and inexpensive consumer receivers, each with an aerial, decoder and controls to switch on or off selected circuits and provide information on price, time period, etc., actable on by consumers. Most systems are one-way but, economically justifiable for large consumers, interactive return transmitters can be installed. UK receivers costing little more than time-of-use variable-rate meters have:

(a) Time clock updated by broadcast signals
(b) Program store adjustable in advance by broadcast messages
(c) Immediate circuit switching by general broadcast messages.

Demand forecasting, management and reliability 59

Utility load shape objectives		Residential	Examples of customer options Commercial	Industrial
Peaking clipping, or reduction of load during peak periods, is generally achieved by directly controlling customers' appliances. This direct control can be used to reduce capacity requirements, operating costs, and dependence on critical fuels.		• Accept direct control of air conditioners	• Accept direct control of water heaters	• Subscribe to interruptible rate
Valley filling, or building load during off-peak periods, is particularly desirable when the long-run incremental cost is less than the average price of electricity. Adding properly priced off-peak load under those circumstances can decrease the average price.		• Use off-peak water heating	• Store hot water to augment space heating	• Add nighttime operations
Load shifting, which accomplishes many of the goals of both peak clipping and valley filling, involves shifting load from on-peak to off-peak periods, allowing the most efficient use of capacity.		• Subscribe to time-of-use rates	• Install cool-storage equipment	• Shift operations from daytime to night-time
Strategic conservation involves a reduction in sales, often including a change in the pattern of use. The utility planner must consider what conservation actions would occur naturally and then evaluate the cost-effectiveness of utility programs intended to accelerate or stimulate conservation actions		• Supplement home insulation	• Reduce lighting use	• Install more efficient processes
Strategic load growth, a targeted increase in sales, may involve increased market share of loads that are or can be served by competing fuels, as well as development of new markets. In the future, load growth will include greater electrification — electric vehicles, automation, and industrial process heating.		• Switch from gas to electric water heating	• Install heat pumps	• Convert from gas to electric process heating

Figure 3.3 Load shape and demand side alternatives (Source: CORY, B.J. 'Load management by direct control', internal paper, Department of Electrical Engineering, Imperial College, 1985, Figure 1. Reproduced by permission of the College)

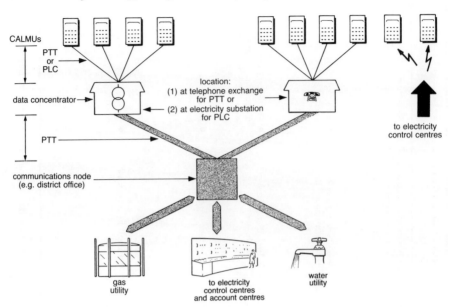

Figure 3.4 *CALMS communications system (Source: CORY, B.J. 'Load management by direct control', internal paper, Department of Electrical Engineering, Imperial College, 1985, Figure 2. Reproduced by permission of the College)*

This system's advantage is central control equipment needing installation only once, covering large areas using existing radio facilities without giving or receiving interference.

Using the second communications mode above (i.e. telephones or pilot wires), DM for large consumers employs techniques borrowed from electricity networks supervisory control, many industries with on-site generation having direct wire links available anyway. For small consumers one-way communication is normal, but two-way communication is possible with incentives for consumers installing the necessary equipment. Most telephone utilities provide data transmission services readily employable for DM. Cable television provides similar facilities, as do satellites.

Also using telephones is the UK Credit and Load Management System (CALMS), Figure 3.4, which can be installed in residences which, in full system, have two-way communication with utilities. One trial installation of 300 units is fully documented, and other trials somewhat less so [9]. For these, four-level time-of-day tariffs were on circuit, with two-level tariffs for interruptable loads and night rates for time-switched supplies. Using the third communications means (DM communicating mainsborne over electricity systems by superimposing higher frequencies than the 50 or 60 Hz mains) is well tried under 'ripple control' schemes, covering wide areas applied to the high voltage sides of step-down distribution transformers, through which it does not pass.

Ripple control injection equipment exploits invested capital by extending the number and versatility of receivers, providing many DM functions through a coded range. It is basically one-way because of the injection power levels needed and the limited bandwidth, large amounts of data or codes not being readily transmittable. Modern solid state techniques give high reliability at reasonable cost and mainsborne, especially in Europe and Australia, is being extended. A new development superimposes frequency ranges on the mains between 50 kHz and 150 kHz using spread-spectrum techniques. Only low power is required, so two-way low-cost signalling between low voltage secondaries of supply transformers and network consumers becomes possible. Because signals will not pass through high voltage transformers, only local communication is obtained, needing telephone or radio links from substation to area control.

3B.2.4 Demand controllers

Modern communication techniques enable separate circuits in commercial or residential premises to be individually controlled, requiring DM wiring during installation or inexpensive modifications to existing wiring. Mainsborne signalling inside residential or commercial premises is an obvious choice (many inexpensive controllers marketed) since signal power is low with short signalling distances, but there can be interference through neighbouring systems. Superimposed frequencies and coded signals are easy to implement but, if a mains input series-connected blocking filter is required, the cost can be prohibitive for widescale use. At appliances, simple plug-in switch units can be employed, switching circuits on and off on receipt of preset codes from local transmitters; these may have individual stored programs, adjusting loads upon receipt of systemwide DM signals, or be arranged to work at various load levels (e.g. level one reduces load by 10%, level two reduces load by a further 10%, level three is for minimum load only). Restoration is at given rates or preset time periods. Rather than have fixed percentage reductions required on command, industrial and large commercial consumers need one to two hours curtailment notice for a firm level, decided initially by consumers after utility advice, usually renegotiable annually.

For reducing demand, consumers obtain rebates on monthly bills dependent on curtailments agreed to. Even for residences with DM, but especially for industry and commerce, consumer–utility relations improve with DM rebates. Another technique is automatic curtailment of loads when frequency falls by (say) 0.15 Hz. In small systems this provides emergency load shedding for systems under stability difficulties.

3B.2.5 Available options

For maximum benefit DM needs integrating with hierarchically designed utility control. Fortunately, modern technology allows this through add-on microprocessors with access to utility databases. DM apparatus is still mainly for peak clipping, enabling better resources utilisation at cheaper cost than installing extra supply plant for peak. It is most effectively applicable initially in industrial or large commercial installations because of large load blocks controllable from single points. For residential or small commercial cases, cheap, reliable DM requires considerable standardisation of everything, bearing in mind that one-way DM communication achieves useful demand clipping plus additional information, e.g. price and curtailment warnings. Also two-way communication, enabling consumer response to DM signals to be assessed, remote energy meter reading and billing applied, is desirable, but may not always be economic.

3B.3 DM by contract

DM applied to both peak and off-peak provides useful 'benchmarks' for measuring efficient use of electricity systems and fuel purchases. Before marginal cost pricing became normal, there was only discouragement of peak usage, mostly by direct pricing means. Another related means was and remains by contract, using special, often complicated, commercial agreements between utilities and largish industrial or commercial consumers (e.g. steel, chemical works and office estates), offering DM consumers lower off-peak tariffs. Successful contract agreements should preferably:

(a) Clearly describe agreement details
(b) Indicate anticipated advantages and disadvantages to both utility and consumer
(c) Quantify and evaluate cost-effectiveness of agreement to both utility and consumer
(d) Set up procedures for modifying the agreement.

Such agreements will succeed if utilities and consumers have DM action plans, including improving energy efficiency and energy conservation (Chapter 4). Alternatively, DM by contract is through regulatory bodies imposing working patterns, staggering industrial production or business activity thus smoothing out electricity demand, and insisting new industrial or business premises have emergency autogeneration or cogeneration, directly operable at peak. However, in North America and elsewhere, utilities must justify to the regulator expenditure on all new plant, quoting all alternatives including DM, investment and borrowing always to be minimised. DM by contract is investment-releasing enabling existing peak capacity to cope longer with load growth, and is pollution free and environmentally friendly, these factors being particularly telling if nuclear plant is its alternative. With the increased privatisation and competition of the 1990s, DM commercial contracts between generators, main, auto and cogenerators, network utilities and large consumers became crucial.

3B.4 DM by pricing

3B.4.1 Use of pricing

A single tariff designed only to recover costs can control demand below the required level, even for subsidised electricity. Sophisticated tariffs provide differentiated rates for time-of-use, by day or by season, reflecting supply costs of those times. One simple form is offering lower rates off-peak, providing a powerful DM technique, moving consumption towards lower electricity production cost times, mutually benefiting suppliers and consumers alike. Logical extensions are variations in consumers' service, using special tariffs for varying consumption patterns and providing for utility intervention to reduce consumption. This DM technique, originally developed for large industrial or commercial consumers in return for tariff advantages, is required at short notice to reduce consumption for typically hourly periods. Technology costs have fallen while energy costs have risen, thereby enhancing DM's case. Peak and off-peak pricing is pursued further in Chapter 6.

3B.4.2 Price incentives

DM price incentives allow consumers to vary demands yet obtain benefit, best examples being illustrated in wholesale Bulk Supply Tariffs (BST) which normally have peak demand (kW) and energy (kWh) charged separately, depending on time-of-day or period. This causes large industries or commercial organisations (e.g. hotels, offices) to control demand by installing their own DM schemes, the necessary equipments being available (e.g. electronic meters, timers and processors arranging for optimum electricity usage including energy conservation). Properly constituted higher prices paid by industry through BSTs has led to improvements in fuel use efficiency but greater savings are still possible in electroheat and motive power capital. With cheaper, sophisticated electronic controllers and improved communication, DM price incentives is spreadable in the 1990s to medium domestic and commercial consumers, producing large demand impacts at certain periods and reducing future costs, e.g. DM by CALMS units mentioned earlier. Technology will allow householders to select various prices per kWh levels for selected appliances (e.g. freezers, water heaters, dishwashers, etc.) to be switched off or on. This works much better in real-time rather than pricing prescribed well in advance (Chapter 6). However, studies are needed dealing with:

(a) Overcoming any remoteness of the electricity sector, especially with respect to placing consumers on automatic meter reading, probably with gas, water, telecoms utilities sharing both the cost and the facility
(b) Increasing utility responsiveness to individual consumers' problems
(c) The implications of domestic and commercial interruptible tariffs on DM by price
(d) DM changing billing frequency and speed, and collecting procedures.

3B.5 Consumer response

3B.5.1 Controlled utilisation

In the 1990s electricity will contribute an increasing proportion of energy demands, but it is prudent to discourage wasteful usage [10]. Controlled, efficient electricity growth by DM is becoming generally accepted for industrial and commercial sectors, and even for residences. DM always results in using generation more efficiently, and can result in lower electricity costs, to be passed on to those consumers assisting DM. Consumers are not motivated unless deliberate action, preferably by utilities, is taken and even then response is slow, because objectives are dependent upon the individual decisions of thousands of consumers, related eventually to new homes, offices, factories and machinery,

64 *Demand forecasting, management and reliability*

when existing equipment reaches its useful life's end. Figure 3.5 shows consumer response in supply–demand interrelations, involving direct connections with planning, operations, tariff design and finance. The process is circular with feedback loops not shown to preserve clarity, e.g. between load shapes and tariff design, between planning and finance, between operation and consumer response.

Figure 3.6 illustrates the effects on daily demand curves in broad load-modification categories, aimed at by supply-demand interrelationships. The figure involves:

(a) Conservation; reduced demand throughout the daily demand against time curve
(b) Substitution to other fuels from electricity at times of daily peak
(c) Interruption; some supplies switched off at peak
(d) Storage; some electricity previously used at peak used off-peak
(e) Peak delay; some energy previously at peak transferred to peak 'shoulders'

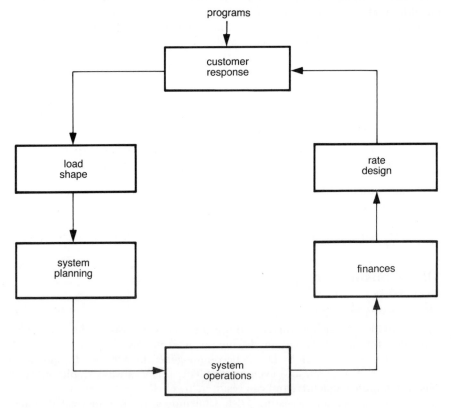

Figure 3.5 *Interrelationships (Source: HOSE, J.F. 'Demand side planning; a practical perspective', IEEE Power Engineering Society, Winter Meeting New York, 1986. Reproduced by permission of the IEEE)*

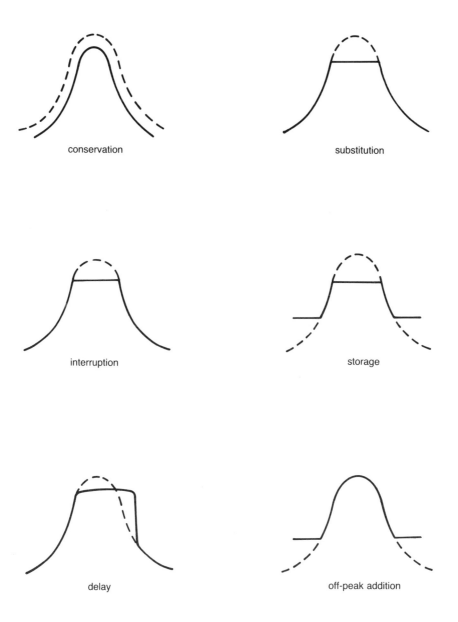

Figure 3.6 *Load modification categories – – – – unmodified, ——— modified. (Source: HOSE, J.F. 'Demand side planning; a practical perspective', IEEE Power Engineering Society, Winter Meeting New York, 1986. Reproduced by permission of the IEEE)*

66 Demand forecasting, management and reliability

(f) Off-peak addition without peak reduction; additional energy used off-peak.

3B.5.2 Case study 1

On 22 August and 11 September 1985 some USA consumers [11] were subjected to three, 15-minute operations of load control equipment, the times of operation being:

No. 1	3.15 pm to 3.30 pm off
No. 2	4.30 pm to 4.45 pm off
No. 3	5.45 pm to 6.00 pm off

Figure 3.7 shows the control schedule.

The temperatures recorded on 22 August and 11 September are in Table 3.2. The area temperature at test time was in the high 80s and the low 90s, but some local showers caused temperatures to moderate in some areas, Table 3.3. During this test, demand data were recorded as usual, plus special efforts to record selected substation transformer loads of high saturation prime time equipment, Table 3.4.

The test purpose was to:

(a) Check DM equipment was functioning properly
(b) Observe test impact on system
(c) Observe test impact on distribution substations
(d) Determine average consumer demand reduction; compare with previous data
(e) Present graphic results of test impact on distribution and other systems
(f) Confirm validity of estimating criteria used.

Conclusions reached were:

(a) Based on demand reduction estimates from previous monitoring and average DM reduction

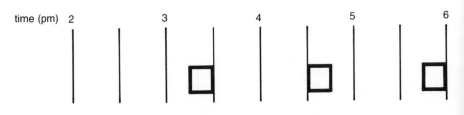

Figure 3.7 *Experimental notch test (Source: STEPHENS, R. E. 'Notch test results', IEEE Winter Power Meeting, 1986, New York. Reproduced by permission of the IEEE)*

Table 3.2 *Temperature data (Source: STEPHENS, R.E. 'Notch test results', IEEE Winter Power Meeting, 1986, New York, USA, Attachment 1. Reproduced by permission of the IEEE)*

Hour	Temperature (deg. F) 22 Aug.	11 Sept.
1	79	80
2	79	79
3	79	78
4	78	77
5	77	74
6	77	76
7	77	78
8	77	79
9	82	82
10	86	86
11	88	87
12	90	87
13	89	86
14	88	89
15	91	91
16	91	90
17	88	90
18	87	84
19	86	84
20	84	83
21	78	78
22	79	78
23	78	80
24	77	79

(b) Load reduction was approximately 88% of expected, 85 to 90% of DM equipment being effective
(c) Such annual tests provide comparative determination if equipment continues effective; valid summer bases were established by the two tests
(d) Impact of tests seen on system load (Figures 3.8 and 3.9), circuits and substation transformer load recordings (Figure 3.10)
(e) Average load reduction for consumers fed from the 12 substation transformers was about 2 kW per consumer, (Table 3.3); 5710 prime time customers in this sample with load control saturations:
   Water heater   98·4%
   A/C short   57·7%
   A/C extended   21·9%
   (Total A/C)   79·6%
   Pool pump   20·8%

Seventeen commercial or industrial consumers were in this sample with load control saturations primarily:
   Water heater   35%

A/C short   100%

These 17 DM commercial or industrial consumers were served by substations in the test sample. The June 1985 peak period average for peak demand for the 90 monitored commercial and industrial consumers was 20.9 kW (valid sample of 61 consumers). Assuming these 17 were average, aggregate commercial or industrial demand reduction is 355 kW. Individual consumers peak hour check-back of electricity use monitoring showed the 20·9 estimate to be reasonable. Estimated load reduction was based on the recorded number of consumers implemented at the time of the test (Table 3.2) for the sample area using load reduction estimates established from 19, 20, and 21 June 1985, end-use data for water heaters, air conditioners and pool pumps. Sample sizes were:

       Water heaters   161
       Air conditioners   255
       Pool pumps   5

Estimated load reductions = number of controlled water heaters times estimated load reduction per water heater, plus number of controlled air condi-

Table 3.3 *Load reduction per customer (Source: STEPHENS, R.E. 'Notch test results', IEEE Winter Power Meeting, 1986, New York, USA, Attachment 6. Reproduced by permission of the IEEE)*

	30/8/85 Customer count	Average load** reduction total		Reduction* per customer	
		22 Aug	11 Sept	22 Aug	11 Sept
1. S. Seffner W	275	550	750	2·0	2·7
2. S. Seffner E	480	850	1200	1·9	2·5
3. Bloomingdale N	652	1000	1450	1·7	2·2
4. Brandon W	152	250	150	1·8	1·0
5. Brandon E	458	800	900	2·2	2·0
6. Jackson Rd. E	480	1000	850	1·8	1·8
7. Jackson Rd. W	462	1050	900	2·8	2·5
8. Pine Lake N	282	700	550	2·5	2·0
9. Pine Lake S	376	600	750	1·6	2·0
10. Carrollwood E	491	1300	1050	2·6	2·1
11. Carrollwood W	905	1700	1400	1·9	1·5
12. Patterson Rd.	697	1650	1400	2·7	2·0
Total	5710	11 450	11 350		
With 19 C/I customers				2·0	1·98

Caution!
* This load reduction reflects that reduction obtained by interrupting all applicable customers. It is not valid as an estimate of load reduction under normal circumstances since a large portion of controlled air conditioners are cycled.
** The average load reduction was determined by taking an average of the three load reductions experienced immediately after the control was initiated, rounded to the nearest 50 MW.

Table 3.4 *Load data collected for notch test, 22 August and 11 September, 1985 (Source: STEPHENS, R.E. 'Notch test results', IEEE Winter Power Meeting, 1986, New York, USA, Attachment 2.1. Reproduced by permission of the IEEE)*

Data source	Medium	Comments

Patterson Rd. Sub.
  Circuit 13869
    13863
    13864
    13860
	AS&E PLC	

Carrollwood Vil. Sub.
  Circuit 13535
    13538
    13539
    13540
    13541
    13544
	AS&E PLC	

Ehrlich Rd. Sub.
  Circuit 13886
    13889
    13890
	AS&E PLC	Pertinent data lost due to PLC communication problem

Brandon Sub.
  E. Transformer
  W. Transformer
	Sys. Ser. data
	Acq. Sys.	

Bloomingdale Sub. N. Transformer	Sys. Ser. data Acq. Sys.	

Carrollwood Vil. Sub.
  E. Transformer
  W. Transformer
	Sys. Ser. data
	Acq. Sys.
		This data was collected manually by taking as many readings as possible just before, during and just after the notch test control was exercised.

Jackson Rd. Sub.
  E. Transformer
  W. Transformer
	Sys. Ser. data
	Acq. Sys.	

Pine Lake Sub.
  N. Transformer
  S. Transformer
	Sys. Ser. data
	Acq. Sys.	

S. Seffner Sub.
  E. Transformer
  S. Transformer
	Sys. Ser. data
	Acq. Sys.	

Individual Prime Time Monitoring Customer	Mag. tape and PLC Load Survey	Approx. 300 customers
Individual C/I Load Control Customer System	Solid state recorder via phone line Sys. Operations Sys. load recorder	Approx. 90 customers

tioners times estimated load reduction per air conditioner, plus number of controlled pool pumps times estimated load reduction per pool pump, plus number of commercial and industrial consumers times estimated average load reduction.

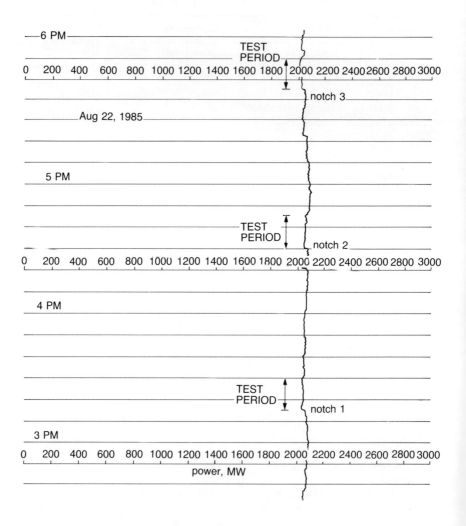

Figure 3.8 *Tampa electric system load (Source: STEPHENS, R.E. 'Notch test results', IEEE Winter Power Meeting, 1986, New York. Reproduced by permission of the IEEE)*

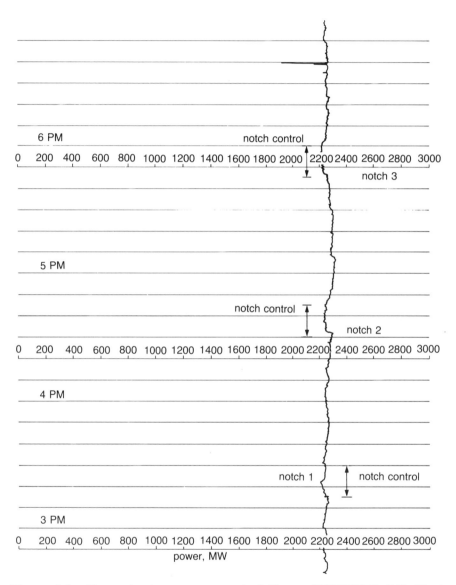

Figure 3.9 *Tampa electric company system load (Source: STEPHENS, R.E. 'Notch test results'. IEEE Winter Power Meeting, 1986, New York. Reproduced by permission of the IEEE)*

Putting these into figures from data already mentioned gives:
Estimated load reductions = 5539 (0·26) + 4854(2·07) + 1149(0·92)
 + 17(20·0)
 = 1440·1 + 10 047·8 + 1057·1 + 355·3
 = 12 900·3 kW
 = 12·9 MW

Effectiveness ratio = Perceived value with respect to estimated value of load reduction
 = 11·45/12·9 = 0·89 or 89% for 22 August 1985 loading
 = 11·35/12·9 = 0·88 or 88% for 11 September 1985 loading

3B.5.3 *Case study 2*

Some UK field trials for the three DM techniques by apparatus [8] evaluated technical performance and quantified utility and consumer benefits, including: generation and distribution cost savings by load curve reshaping; improving cash flows from more frequent billing; provision of daily information on

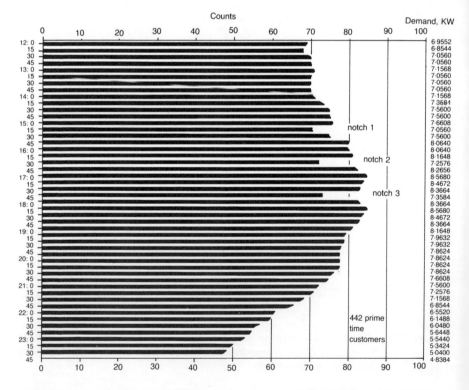

Figure 3.10 *ASEP demand profile report (Source: STEPHENS, R.E. 'Notch test results'. IEEE Winter Power Meeting, 1986, New York. Reproduced by permission of the IEEE)*

electricity costs and usage; and facility to introduce more flexible tariffs. The radio teleswitching data broadcasts were superimposed on 200 kHz long wave transmission using phase modulation. A data '1' was indicated by a phase advance of 25 degrees for 20 ms, followed by a phase retard of 25 degrees for 20 ms; a data '0' was indicated by a phase retard of 25 degrees for 20 ms followed by a phase advance of 25 degrees for 20 ms. Thus the data spectrum did not overlap broadcast material, data being filtered so that overall data channel spectrum-shaping was the '100% cosine roll-off' shape with a half-amplitude point of 25 Hz. Data was sent continuously in frames of 50 bits, each frame taking two seconds to broadcast and consisting of a synchronising prefix, a 4 bit application code, a 32 bit message and a 13 bit check word. Each distribution utility was allocated a unique code and each teleswitch would only receive codes initiated by the owning utility. No one code controlled more than 100 MW. Approximately 1700 prototype radio teleswitches obtained from one manufacturer were used by most utilities to gain experience and also to examine technical performance and consumer reaction. Samples of consumers ranging from small numbers of utility employees to nearly 200 other domestic consumers on two-rate tariffs were recruited, with radio teleswitches used to replace existing mechanical time-switches. Opportunity was taken to experiment with complex tariffs, e.g. varying starting times of afternoon two-hour boosts of an eight-hour + two-hour off-peak tariff. Other experiments included a multi-rate tariff with three rates, a peak rate for winter daytime weekdays except holidays, a normal rate for evenings and weekends through the year and weekdays outside winter, and a low night rate. The result can be perused in detail in reference [8]. The mainsborne telecontrol scheme [12] tested was a joint development by UK electricity, gas, and water utilities, comprising:

(a) The central controller in the electricity distribution supply substation
(b) Home unit receiving signals from water, gas and electricity meters at the electricity meter positions
(c) Consumer display units.

Illustrative problems arose from branched network and variable complex consumer load impedances. Any system must be operable with varying frequency responses, impedances, and attenuations, surviving large impulses and operating successfully with small impulses and with consumer interference at varying frequencies. The spread spectrum technique meets these requirements well and was used in the trials, data being spread over a frequency band by using code sequences of 1024 bits, transmitted at data rates of 200 bits per second. Since time between individual pulses and pulse duration varies, information was spread over a wide frequency band. At the receiver, the pulse pattern was reference-correlated, precautions taken allowing for differences in synchronisation between transmitter and receiver clocks. The coding generation and decoding by correlation at the receiver was carried out by microprocessors. Since the main purpose of the trials was testing mains reliability as a communication medium, consumers continued to be billed normally and hence no financial inducement to consumers to participate in the trials was necessary. Evening presentations were held for householders followed up by house-to-house surveys when agreement to participate was confirmed. During trials, in addition to network use for operation by utilities, home units were polled in turn throughout the 24 hours, using standard messages seeking status information, failure to obtain the information being logged and results plotted, showing:

(a) Percentage of home units successfully communicated with at various times throughout each day
(b) Variations between days
(c) Variations in success rate between different home units over 24-hour periods.

Performance during June 1984 confirmed that reliable communication was obtainable to most home units at all times of day and to all home units at some time during the 24-hour periods, there being variations in performance over the 24 hours depending also on location.

For the CALMS trials [8], three locations in different areas were chosen being selected convenient for utilities, consumers and telecoms. For each site, 100 consumers were selected as a representative sample; what was required were domestic consumers with telephones, being unlikely to move house. Other factors considered were: present tariff; appliance ownership; previous consumption patterns; willingness to install new meters (e.g. for water); and sufficient space for a CALM meter. Every transaction between utility control computers and CALMS meters was logged for later analysis. Load profiles plus equipment and communication reliability were studied, plus comparisons with other load research data. Engineering information was also studied. Some conclusions drawn were: consumer reaction, established by responses to questionnaires, was extremely favourable and the CALMS meter was well able to convey energy cost messages and other information; tariff structures were well understood by consumers, who appeared willing to react to new ones; little inconvenience was reported from load management carried out; trial switches were not proving to be entirely satisfactory; but CALMS, communications schemes, and central computer systems were performing well.

3B.6 Reliability

3B.6.1 1990s approach

In the 1990s reliability will be treated increasingly from the demand side rather than from the traditional supply side. This is because:

(a) Widespread, accurate information is now available on outage costs
(b) DM schemes enable suppliers to trim demand at short notice to almost arbitrary values
(c) Dynamic pricing accelerates (b).

Reliability will ultimately become part of DM's remit by, in the optimum way, quickly making sure that demand and supply always equate. Unplanned blackouts [13] produce large outage costs, particularly to industry, products being spoiled and normal production being interrupted, resulting in opportunity costs in idle capital and labour during outages and following restart. Some, but not all, lost value-added is recoverable by using capacity more

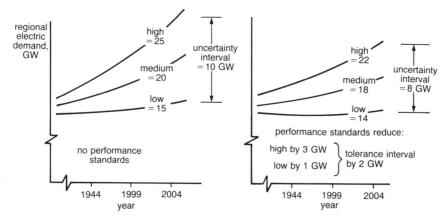

Figure 3.11 *A simple numerical example of uncertainty reduction (Source: FORD, A. and GENZER, J. 'The impact of performance standards on the uncertainty of the Pacific North West System' report to the Bonneville Power Administration, 1988, USA. Reproduced by permission of Bonneville Power Administration)*

intensively during normal hours, and/or operating overtime. Domestic consumers having no marketable outputs require an approach of relationships between incremental wage rates and leisure value foregone due to electricity outages. Survey techniques provide additional information on residential users' willingness to pay to avoid outages, or accept compensation for inconveniences and costs incurred during outages.

3B.6.2 *Case study I*

A detailed study of uncertainty [14] quantified the extent to which set standards of reliability reduce uncertainty. Figure 3.11 shows plausible situations where impact of reliability standards is easy to understand, the left side showing three electricity growths with no reliability standards in effect. With high and low projections of 25 GW and 15 GW respectively, the uncertainty interval is 10 GW. The right side shows three projections with standards which reduce demand in low growth by 1 GW because of new buildings demand reductions. The medium growth has twice the new buildings so has twice the reduction while the high growth case has three times the reduction because it has three times the new buildings. Standards work against uncertainty in economic growth, producing greater savings as economic growth increases; here, high growth incurs an extra 2 GW of savings over low growth, causing a 2 GW reduction in uncertainty. These figures are given for reference to help understand relative magnitudes and plausibility of results. The case study showed:

(a) Both demand and tariff uncertainty grow large over time; Figure 3.12 shows regional uncertainty of growth becoming 50% of medium forecast
(b) The utility bears a disproportionate share of uncertainty in its role of power broker, facing uncertainty that grows to over 80% of medium projection

(c) Efficiency standards reduce long-term uncertainty in loads and tariffs, sometimes dramatically
(d) Money values of reduced uncertainty are important when evaluating tariff's impact but unimportant for revenues or consumer costs.

Regional or district demand uncertainty in Figure 3.12 and Figure 3.13 showed tolerance intervals of 10 300 MW by the year 2004, 50% of the mean projection, a result surprisingly close to simple illustration in Figure 3.11; having standards for reliability cut this uncertainty by 870 MW or about 8%. However, the overall utility situation is different, demand uncertainty being 7600 MW in 2004, over 80% of mean projection. Reliability standards cut this uncertainty by 800 MW or a 24% reduction. The dramatic difference between regional and utility results in Figures 3.12 and 3.13 show how standards cut mean projections of demand (horizontal axis) is in the 45 degree line. The main conclusions are that reliability standards are more effective in cutting utility demand uncertainty than regional or district uncertainty.

Uncertainty impacts change with changes in the design of efficiency standards themselves:

(a) Less aggressive standards have less effect on reducing uncertainty, and

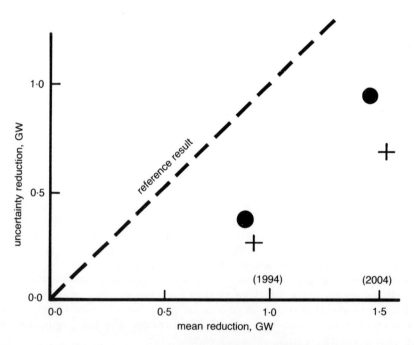

Figure 3.12 *Standards impact on regional demand.* ● *final viewpoint,* + *initial example. (Source: FORD, A. and GENZER, J. 'The impact of performance standards on the uncertainty of the Pacific North West System' report to the Bonneville Power Administration, 1988, USA. Reproduced by permission of Bonneville Power Administration)*

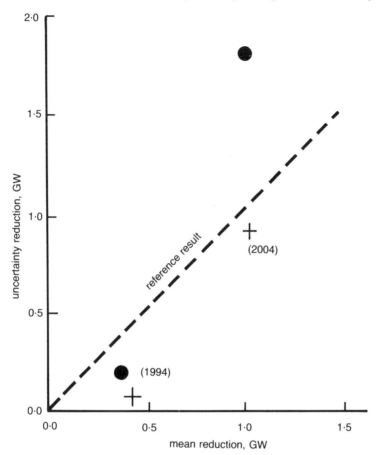

Figure 3.13 *Standards impact on Bonneville load.* ● *final viewpoint,* + *initial example. (Source: FORD, A. and GENZER, J. 'The impact of performance standards on the uncertainty of the Pacific North West System' report to the Bonneville Power Administration, 1988, USA. Reproduced by permission of Bonneville Power Administration)*

vice versa, ability of standards to reduce demand and uncertainty being fundamentally related.
(b) All-fuel standards are more effective at reducing uncertainty than electricity-only standards.
(c) Uncertainty to compliance has no significant effect on the ability of standards to reduce uncertainty.

Specifically interesting is whether greater downside or upside demand variations are likely, important because contingency measures against downside variation are different from those for upside variation. An overall region or district has greater potential for downside variation due to possible loss of load from (say) aluminium smelters. When looking at a utility's uncertainty, however, much greater upside variation exists, due to the way loads are actually placed on the utility. It can be concluded that standards would cut the wholesale tariff uncertainty in the year 2000 by 0·25 US cents/kWh, corresponding to a 22% reduction in uncertainty for the utility's average wholesale consumer.

3B.6.3 Case study II

Assessment was made [9] of the likely effects of DM on domestic and commercial consumer classes, on the peak of the UK generation and transmission system over the years 1990 to 2010. The calculations started for the domestic consumer class with actual and projected unrestricted sales given in Table 3.5 showing sales for each appliance type considered suitable for one of

Table 3.5 *Actual and estimated sales of electricity in 1979/80 and 2000/1 (CEGB Scenario C) (Source: CORY, B.J. 'Proof of evidence to Sizewell B enquiry; load management and peak electricity demand'. Department of Electrical Engineering, Imperial College, 1986, Table I. Reproduced by permission from the College)*

	Sales TWh	
	1979/80	2000/1
Domestic (unrestricted sales) (a)		
Refrigerators	4·0	2·5
Fridge-freezers	1·9	4·1
Freezers	5·0	7·4
Cookers	7·7	6·9
Space heating	7·6	6·2
Water heating	12·4	2·6
Other	23·8	28·2
	62·4	57·9
Restricted sales (off peak)	11·2	20·3
Domestic total (England & Wales)	73·6	78·2
Commercial sales (restricted and unrestricted) (b)		
Space and water heating	12·9	19·7
Cooking and catering	6·9	9·6
Lighting	21·2	22·2
Airconditioning	3·8	5·3
Public lighting	2·2	2·2
Other	3·9	4·5
Commercial total (UK)	50·9	63·5
of which	4·4 (8.6%) is off peak	9·4 (14·8%) is off peak
Commercial sales (England & Wales)	41·9	52·7

Notes: (a) From Table 59 CEGB/P5 (C.M. Davies)
       (b) From Tables 89, 90 and 91—CEGB/P5 (C. M. Davies) with additional data supplied by CEGB.

Table 3.6 *Contributions of domestic loads to demand at time of peak with estimated effect of LM, 1979/80 (Source: CORY, B.J. 'Proof of evidence to Sizewell B enquiry; load management and peak electricity demand'. Department of Electrical Engineering, Imperial College, 1986, Table II. Reproduced by permission from the College)*

Domestic load component	% LM possible to apply (a)	Peak demand (GW)			
		1700 hrs		1730 hrs	
		Without LM (b)	With LM	Without LM (b)	With LM
Refrigerators	90	0·40	0·04	0·40	0·04
Fridge-freezers	90	0·20	0·02	0·20	0·02
Freezers	90	0·50	0·05	0·50	0·05
Cooking	0	3·00	3·00	2·80	2·80
Space heating	20	3·40	2·72	3·50	2·80
Water heating	100	2·20	0·00	2·90	0·00
Other	10	5·60	5·04	7·60	6·84
Totals		15·30	10·87	17·90	12·55
Peak load lopped		29%		30%	

Notes: (a) These are estimated percentages of load which can be delayed at peak for up to one hour;
(b) Data obtained from CEGB in response to letter from CPRE dated 19 July 1982

the DM schemes described earlier. Table 3.6 shows estimated peak contributions in 1979/80 of various domestic load types at two half-hour peak periods. Also shown are estimated percentage demands delayable for up to one hour by DM for each load type, ranging from 90% for cooling devices with considerable thermal time lag to 10% for other loads including laundry appliances and dishwashers, and lighting. Table 3.7 shows similar estimates for domestic contributions to peak half-hours in the year 2000 based on extrapolation from unrestricted sales of Table 3.6. From Tables 3.6 and 3.7 it is concluded that DM could 'lop' the domestic peak by between 21% and 30%; for planning purposes say 20%. A similar estimation was done for commercial loads with demand breakdown into components at peak not available. Table 3.8 gives commercial sales for 1979/80 for both restricted and unrestricted use. From these an average demand over the year was calculated and to this was applied possible percentages of DM which could delay the peak for up to one hour. For space or water heating and air conditioning, 50% of this load was assumed to be peak lopped. Allowing for institutions and educational establishments where any load restriction causes hardship, then 50% restriction is acceptable. Table 3.8 allows estimates that a 15.6% reduction in average demand is achievable; for planning purposes 15% for commercial loads. This is conservative in that for 1979/80 actual figures indicate a contribution to peak of 11 GW by commercial loads, compared to 5·6 GW of average demand shown in Table 3.8, a potential for DM of around 25%.

Having estimated plausible DM effects when first installed, next is required its future progress in penetrating the domestic and commercial markets. Estimates suggest 95% coverage of domestic consumers within ten years of commencement of a vigorous policy of DM. In order to arrive at a meaningful assessment it is first necessary to estimate the contribution at peak made by domestic and commercial sectors from 1990 until the 95% market penetration at about 2010. Table 3.9 shows possible percentage contributions made by these sectors without DM at the two peak periods by extrapolating 1979/80 figures through the assumed figures to 2000/1. Contributions to peak in later years are proportional to anticipated consumptions given in Table 3.6. An average percentage contribution for both sectors is considered since the system peak is likely to occur any time between 1700 and 1800 hours. Finally, the effect of DM on the peak in the domestic sector is calculated in Table 3.10 from 1990 to 2010 spanning the period DM is installed, DM effects related directly to utility demand estimates. Peak demand could be reduced by 2·9 GW in 2010/1 through DM applied to the domestic sector alone. The figure for the commercial sector alone is shown in Table 3.11, with less reduction of peak than in the domestic case but by 2010/3 a reduction of 2·2 GW being possible. Finally

Table 3.7 *Contributions of domestic loads to demand at time of peak including estimated effects of LM, year 2000/1 (Scenario C) (Source: CORY, B.J. 'Proof of evidence to Sizewell B enquiry; load management and peak electricity demand'. Department of Electrical Engineering, Imperial College, 1986, Table III. Reproduced by permission from the College)*

Domestic load component	% LM possible to apply (b)	Peak demand (GW)			
		1700 hrs		1730 hrs	
		Without LM (a)	With LM	Without LM (a)	With LM
Refrigerators	90	0·25	0·025	0·25	0·025
Fridge-freezers	90	0·43	0·043	0·43	0·043
Freezers	90	0·74	0·074	0·74	0·074
Cooking	0	2·69	2·690	2·51	2·510
Space heating	20	2·75	2·200	2·83	2·264
Water heating	100	0·46	0·000	0·61	0·000
Other	10	6·63	5·967	9·00	8·100
Totals		13·95	11·000	16·37	13·02
Peak load lopped		21·1%		20·5%	

Notes: (a) The peak unrestricted demand for 2000/1 is estimated from the unrestricted sales given in Table 3.5; the ratio between unrestricted sales and contributions of domestic load to peak demand is assumed to be the same in 2000/1 as in 1979/80.
(b) This is the estimated percentage of load which can be delayed for up to one hour at peak.

Demand forecasting, management and reliability 81

Table 3.8 *Commercial demand 1979/80 and 2000/1 (Scenario C) by components and estimated effect of LM (UK) figures (Source: CORY, B.J. 'Proof of evidence to Sizewell B enquiry; load management and peak electricity demand', Department of Electrical Engineering, Imperial College, 1986, Table IV. Reproduced by permission from the College)*

Commercial load component	% LM possible (c)	1979/80 Average demand (a) GW	1979/80 Peak demand before LM (b) GW	1979/80 Peak demand after LM GW	2000/1 Average demand (a) GW	2000/1 Peak demand before LM GW	2000/1 Peak demand after LM GW
Space & water heating	50	1·473	3·093	1·546	2·249	4·723	2·361
Cooking & catering	10	0·788	1·655	1·489	1·096	2·301	2·071
Lighting	0	2·420	5·082	5·082	2·534	5·322	5·322
Air conditioning	50	0·434	0·911	0·456	0·605	1·271	0·635
Public lighting	0	0·251	0·527	0·527	0·251	0·527	0·527
Other	10	0·445	0·934	0·841	0·514	1·079	0·971
Totals		5·611	12·202	9·961	7·249	15·223	11·887
Reduction in peak demand by LM			18·4%			21·9%	

Notes: (a) Calculated by using total sales from Table 3.5 and dividing by 8760 hours.
(b) Obtained by multiplying average demand by factor of 2·1 as taken by M. Barrett CPRE/P/3. This factor is assumed to account for the proportion of off-peak load given in Table 3.5.
(c) This is estimated percentage of load which could be delayed by up to 1 hour at peak.

Table 3.12 shows combined effects of DM on domestic plus commercial sectors on peak demand 1990/1 and 2010/1. Significant savings in peak system capacity amounting to 4.9 GW are feasible.

3B.7 Commentary summary

The management of demand by consumers, enabling the capacity of generating, transmission, distribution and utilisation equipment to be controlled, has a direct effect on planning. Consumers should also benefit by overall cheaper electricity. From the national economy viewpoint, investment is minimum if consistent with supplying consumers' needs, as revealed by their willingness to pay for the quantity, quality and price of electricity. DM is applicable either via supply or demand sides. When peak loads are supplied from expensive-to-operate generation, strong incentives exist from both the national economy and

Table 3.9 *Estimated percentage of contributions to peak demand made by domestic and commercial sectors (Source: CORY, B.J. 'Proof of evidence to Sizewell B enquiry; load management and peak electricity demand', Department of Electrical Engineering, Imperial College, 1986, Table V. Reproduced by permission from the College)*

	Domestic			Commercial		
	17.00 hrs	17.30 hrs	Average of 17.00 & 17.30 hrs	17.00 hrs	17.30 hrs	Average of 17.00 & 17.30 hrs
1979/80	33·0	39·0	36·0	27·0	24·0	25·5
1990/91	31·0	36·5	33·7	28·0	25·0	26·5
1995/96	30·0	35·0	32·5	29·0	26·0	27·5
2000/1	29·0	34·0	31·5	30·0	27·0	28·5
2005/6	28·0	33·0	30·5	31·0	28·0	29·5
2010/1	27·0	32·0	29·5	32·0	29·0	30·5

Figures taken from CEGB data in response to CPRE letter dated 19th July 1982 and 1982–1989 Medium Term Development Plan published July 1982.
Contributions to peak are assumed to vary linearly with time from 1979/80 to 2010/1.

Table 3.10 *Effect of LM on peak demand on CEGB system if installed in domestic premises from 1990 onwards (Scenario C) (Source: CORY, B.J. 'Proof of evidence to Sizewell B enquiry; load management and peak electricity demand', Department of Electrical Engineering, Imperial College, 1986, Table VI. Reproduced by permission from the College)*

Year	LM penetration %	Peak demand GW (c)	
		without LM	with LM
1990/1	0	43·7	43·7
1995/6	25	45·1	44·4
2000/1	50	46·9	45·4
2005/6	75	48·8	46·6
2010/1	95 (max)	50·6	47·7

Notes: (a) Peak demand is assumed to occur between 1700 and 1800 hours. The effect of LM is calculated from average contribution of domestic loads to peak given in Table 3.9.
(b) LM is assumed to be installed at rate of 5% of domestic consumers per year from 1990/1 onwards. Maximum penetration reached is 95% around 2010/1.
(c) Estimated from CEGB/P4/Table 23 and further Demand Projections supplied by CEGB dated 3 March 1983

Table 3.11 *Effect of LM of peak demand of CEGB system if installed in commercial premises from 1990 onwards (Scenario C) (Source: CORY, B.J. 'Proof of evidence to Sizewell B enquiry; load management and peak electricity demand', Department of Electrical Engineering, Imperial College, 1986, Table VII. Reproduced by permission from the College)*

Year	LM penetration %	Peak demand GW (c)	
		without LM	with LM
1990/1	0	43·7	43·7
1995/6	25	45·1	44·6
2000/1	50	46·9	45·9
2005/6	75	48·8	47·2
2010/1	95 (max)	50·6	48·4

Notes: (a) Peak demand is assumed to occur between 1700 and 1800 hours. The effect of LM is calculated from average contribution of commercial loads to peak given in Table 3.9.
(b) LM is assumed to be installed at rate of 5% of commercial consumers per year from 1990/1 onwards.
(c) Estimated from CEGB/P4/Table 23 and further Demand Projections supplied by CEGB dated 3 March 1983

utilities to apply DM in one form or another. In addition, to obtain the best performance from large, modern, thermal generating stations, a steady loading is necessary and DM enables this constant output to be maintained.

Basically there are three methods of applying DM, by:

(1) Direct control of consumer demand
(2) Legal contract
(3) Price

The first and last methods require some special devices in consumers' premises. DM is not concerned with the wholesale broadcast cutting off of consumers, this sometimes being necessary in many developing countries under system catastrophes. DM is designed in fact to avoid such emergency measures. Although DM techniques control or delay mainly peak demand, it can be shown that, with other measures, it produces more effective and efficient use of all existing electricity and the need for new thermal plant can often be delayed by several years. DM also reduces the planning plant margin which traditionally allows for uncertainty, and it can encourage energy efficiency and conservation measures by pricing energy at true cost during high demand periods, as well as providing a mechanism for compensating private generation suppliers for

Table 3.12 *Effect of LM on peak demand of CEGB system if installed in domestic and commercial premises from 1990 onwards (Scenario C) (Source: CORY, B.J. 'Proof of evidence to Sizewell B enquiry; load management and peak electricity demand', Department of Electrical Engineering, Imperial College, 1986, Table VIII. Reproduced by permission from the College)*

Year	LM penetration %	Peak demand GW (c)	
		without LM	with LM
1990/1	0	43·7	43·7
1995/6	25	45·1	43·9
2000/1	50	46·9	44·4
2005/6	75	48·8	45·0
2010/1	95	50·6	45·5

Notes: (a) Peak demand is assumed to occur between 1700 and 1800 hours. The effect of LM is calculated from average contribution of domestic and commercial loads to peak given in Table 3.9.
(b) LM is assumed to be installed at rate of 5% of domestic and commercial consumers per year from 1990/1 onwards. Maximum penetration reached is 95% around 2010/1.
(c) The sum of possible effects due to LM on domestic consumers (Table 3.10) and commercial consumers (Table 3.11).

sudden loss of market. In cost–benefit analysis, outage costs are well enough known to enable their full use in the 1990s with confidence for setting optimum reliability standards, apart from in some developing countries.

3B.8 References and further reading

1 BERRIE, T.W. (1983) 'Power system economics' (Peter Peregrinus) Chapter 4; also Bridger, G. and Winpenny, J. (1984) 'Planning development projects' (Overseas Development Administration UK) p. 74
2 BERRIE, T.W. (1968) 'The economics of system planning in bulk electricity supply' in 'Public enterprise', (edited by R. Turvey), Penguin Modern Economics, p. 173
3 HOLDBERG, P.D. (1987) 'Dealing with energy market uncertainty', *International Association of Energy Economists Newsletter*, **1**, pp. 54–77
4 CHESHIRE, J.H. (1984) 'Energy forecasting: some serious lessons for UK technology policy' (Science Policy Review Unit, University of Sussex, UK)
5 ROBINSON, G. (1990) 'Demand forecasting methods and the need for new nuclear plant' (Surrey Energy Economics Centre Discussion Paper Series)
6 CORY, B.J. (1985) 'Load management by direct control', internal paper (Department of Electrical Engineering, Imperial College, London); also 'Glossary of LM terms' (Edison Electric Institute) report no. EPRI-EURDS, pp. 34–36
7 BERRIE, T.W. (1983) 'Power system economics' (Peter Peregrinus) Chapter 3
8 CALDWELL, J.E., CRAWLEY, D.F., HENSMAN, G.O. and JAMES, A.L. (1985) 'Energy management field trials in UK', *CIRED Conference*, Brighton, pp. 261–265; also Johnson, W.A. and Deveney, T.M. (1985) 'Experimental results from a load management system from large commercial customers', *IEEE PAS-104*, **9**, pp. 2322–

2328; also Beeker, D.L. (1985) 'Load management by direct control: fact or simulation', *IEEE Summer Power Meeting*, 1985, Vancouver, Paper No. 85, SM 477-5

9 HERMAN, H.O., EYRE, B.E., WRIGHT, D.T. and EDWARDSON, S.M. (1982) 'A radio teleswitching system for load management in the UK', *Proceedings of the 4th IEE International Conference on Metering, Apparatus and Tariffs for Electricity Supply*, pp. 40–46; also Peddie, R.A. and Fielden, J.S. (1981) 'Credit and load management systems for an electricity supply utility' *CIRED Conference*, IEE 197 pp. 231–235

10 CORY, B.J. (1982) 'Load management and peak electricity demand, proof of evidence to Sizewell B enquiry' (Department of Electrical Engineering, Imperial College, London); also Platts, J. (1978) 'Britain's experience of electricity marketing directed towards load management', *Energy Conservation and Utilisation of Off-Peak Power Seminar*, USA, 18–27 Sept.

11 STEPHENS, R.E., 'Notch test results', *IEEE Winter Power Meeting*, 1986, New York.

12 CHAPMAN, N.R., EYRE, B.E., GOODWIN, S.J. and WIENER, A. (1982) 'Equipment designs for two-way mainsborne signalling system of energy management', *Proceedings of 4th International Conference on Metering Apparatus and Tariffs for Electricity Supply*, IEE, pp. 108–112

13 MUNASINGHE, M. (1988) 'Optimal planning, supply quality and shortage costs in power system analysis and practical application', *The Energy Journal*, Quarter 3; also same author (1979) 'The economics of power systems reliability and planning' (Johns Hopkins University Press for The World Bank)

14 FORD, A. and GEINZER, J. (1988) 'Adding uncertainty to least-cost planning, a case study of efficiency standards in the North West USA', report to Bonneville Power Administration; also by same authors, (1988) 'The impact of performance standards on the uncertainty of the Pacific Northwest Electric System', in report to Bonneville Power Administration; also Ford, M. and Nail, P. (1987) Bonneville's conservation policy analysis models', *Energy Policy*, April 1987

Chapter 4
Efficiency, conservation and the environment

4A EFFICIENCY

4A.1 Introduction

4A.1.1 *Acceptable losses*

Concerns remaining in the 1990s about fuel availability and price highlight the importance of improving energy efficiency and conservation in energy producers, transporters, distributors and users, e.g. by electricity loss reduction (Figure 4.1). Generation losses decline with improved thermal efficiency and reduced station usage, by:

(a) Using improved technologies, e.g. combined cycle and combined heat and power generators
(b) Replacing old boilers
(c) Uprating thermal generators
(d) Using more efficient hydroplant
(e) Replacing old turbines.

Acceptable generation losses are of the order of a few per cent and average transmission and distribution losses not over 10% of gross generation (economic levels 1% and 5%, respectively), not the after-theft 20% of some developing countries. To illustrate this point [1], India's annual electricity production in 1978 was 10^5 GWh: assuming US cent 5 per kWh, loss reduction from 20% to 10% of generation would have meant annual savings of US$0·5 billion with just a small expenditure on additional hardware. In developed countries, lower percentage losses with higher electricity throughputs still allow scope for savings. Losses are high because electricity prices often are kept below supply costs (e.g. during fuel price rises), leading to reduced investment and maintenance, both effects worsened by low prices overstimulating demand. Generation and transmission shortages lead to widespread outages; this is not so for distribution. Poor voltage and high utilisation losses are not spectacularly apparent like blackouts. New plant is visible, loss reducing improvements not so. Also, in the past, rules for comparing increased investment against reduced losses assumed 1960 ratios of costs between fuels and metals (Figure 4.2). In the 1990s, using additional copper or aluminium or steel in order to reduce expensive fuel losses is economic.

Efficiency, conservation and the environment 87

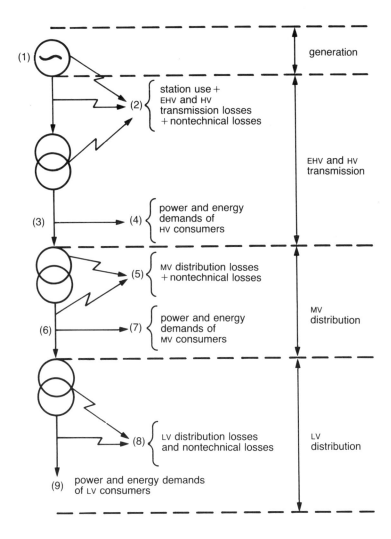

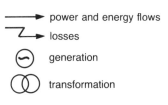

Figure 4.1 *Losses in different parts of the power system (Source: 'Energy efficiency: optimisation of electric power distribution system losses', World Bank Energy Department Paper No. 6, May 1982. Reproduced by permission of the World Bank)*

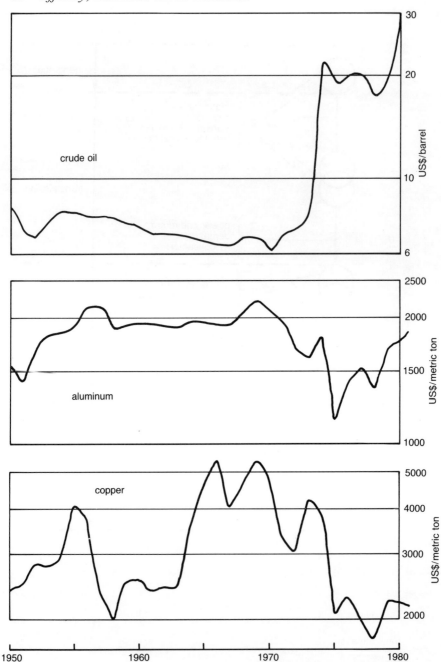

Figure 4.2 *International prices for crude oil, aluminium and copper in constant (1980) US dollars (Source: 'Energy efficiency: optimisation of electric power distribution system losses', World Bank Energy Department Paper No. 6, May 1982. Reproduced by permission of the World Bank)*

4A.1.2 *Studies*

Realising that, for the same reason as above, improving losses is now often more cost-effective than building new capacity (saving capital and increasing utility profits), objectives of studies on losses must:

(a) Identify cost-effective loss optimisation areas
(b) Isolate technical losses from theft, poor billing and collecting
(c) Develop and then test by desk study loss reduction evaluation methods to find optimum loss levels
(d) Develop practical methods for including losses in planning and operating criteria.

In the 1990s, policy issues urgently need studying on overall energy efficiency at all levels of the national energy, utilities and consumers, thereby:

(a) Formulating rules to compare individual strategies and projects
(b) Judging economics of popular schemes; e.g. combined heat and power, combined-cycle plant
(c) Adopting balanced strategies towards efficiency level aimed at different electricity network components
(d) Evaluating techniques from other sectors, especially from entrepreneurs.

With short-term finance becoming paramount in the 1990s, all costs and benefits from improved efficiency must be quantified in financial or accounting terms. This must be done as credibly as possible if improving efficiency as a real alternative to building new projects is to be accepted; in particular calculations should not make exogenous assumptions, as do those of pressure groups following a single theme, e.g. conservationists and environmentalists.

4A.2 Improving production efficiency

4A.2.1 *Thermodynamics*

Thermodynamic's second law declares work to be fully convertible to heat (i.e. 1 kW of work is convertible into 1 kW of heat) but that work obtained from 1 kW of heat cannot exceed an amount which is determined by the temperatures of available heat and its surroundings; the proportion of heat convertible to work, W/Q (i.e. the heat conversion efficiency), is limited to [2]:

$(T - T_o)/T$

where T = temperature of heat source
Q = quantity of heat available
T_o = temperature of surroundings.
Thermodynamic's second law states:

$$\frac{W}{Q} = T - T_o \tag{4.1}$$

Also the maximum work output W_{max} is obtained only when reversible cycle engines are operated between temperature reservoirs T and T_0, when the equality holds and

$$W_{max} = \frac{Q(T - T_o)}{T} \qquad (4.2)$$

100% Q conversion is obtainable if T_o is at absolute zero, when $W_{max} = Q$ at $T_0 =$ zero, purely theoretical, although T equal to infinity is more practical, and conversion processes do exist where work output yields approach 100%, with one work form convertible into another. In practice, however, work is regardable as heat only at very high temperatures.

Equation (4.2) gives measures of heat usefulness at different temperatures, the higher the source temperature (e.g. steam, gases, liquids) the greater the work derived per heat unit. Heat required at 100°C, is obtainable from heat sources at 300°C but not heat sources at 30°C, since heat transfer across finite temperature differences is not reversible, always down and never up the temperature gradient.

4A.2.2 Improving efficiency

It follows that:

(a) Energy at high temperature is more useful than energy at low temperature
(b) Energy conversions always degrade energy quality, all conversions being irreversible
(c) Some conversion irreversibilities are inevitable, e.g. heat transfers across temperature differences required to make things happen
(d) Yet other conversion irreversibilities are avoidable, e.g. inefficient combustion in boilers and poor insulation in heat engines and buildings.

Increasing efficiency means minimising overall requirements of the entire heat to work to energy cycle (e.g. in the capacities of generators, boilers, turbines, condensers, alternators, etc.), taking all that is useful from irreducible minimum energy. Energy not appearing as useful energy means it is non-useful, i.e. the conversion process is to some extent inefficient.

Turbine generation efficiency thus can be improved by [3]:

(a) Ensuring greater differences in temperature between the beginning and the end of heat conversion (equation (4.1)), e.g. by subjecting boilers to higher and steam condensers to lower temperatures; also using higher steam pressures because water becomes steam at progressively higher temperature with increased pressure
(b) Using once-through boilers and fluidised heat beds
(c) Higher gas turbine temperatures
(d) Higher diesel engine operating temperatures
(e) Increasing size of boilers, turbines and alternators, higher temperatures and pressures being more practical with larger units, this all giving economies of scale
(f) Combined-cycle plant, gas turbines or diesels exhausting into steam boilers
(g) Combined heat and electricity generators, with or without district heating.

Steam turboalternator unit thermal efficiencies increased from around 20% in the 1950s to nearly 40% in the 1970s, but no further improvements seem likely. Combined cycle efficiencies are expected to climb to around 50% and combined heat and power to nearly 70%. Such changes imply immense savings in fuel and greater outputs for the same fuel input, improving both energy efficiency and conservation.

4A.2.3 *Station losses*

Generation internal electricity usage and station losses should not be more than a few per cent of output at the most. Low station loss, consistent with safety and efficiency, needs:

(a) Regular maintenance of all auxiliaries
(b) Early repair or replacement of faulty or worn out machinery
(c) Proper operator training
(d) Proper operating procedures
(e) Logging of incidents and maintenance
(f) Regular auxiliaries testing.

In some developing countries extreme shortages of skilled staff, training facilities and funds for local maintenance, repairs and replacements, make it difficult to keep auxiliaries usage efficient, financial help being difficult to obtain for items other than new plant.

4A.3 Combined heat and power

4A.3.1 *Introduction*

Combined heat and power (CHP) takes advantage of thermodynamic's second law, using a heat source between 1000°C and 2000°C and the need for work in the form (a) electricity, and (b) heat, at temperatures of 100°C or below. A further step is to convert energy into heat, then electricity plus some high temperature heat by combustion, then use heat transfer to produce heat at much lower temperatures, heat transfers occurring only at required sites, all combustion conditions being optimised.

4A.3.2 *Industrial CHP*

CHP therefore [4] achieves electricity and heat simultaneously, with greater overall efficiency than each separately. Components are prime movers from which heat rejected is used within heat exchangers, often in boiler format. Component combinations and permutations are many and innovations are always in progress.

In the USA, cogeneration means a 'topping cycle' equivalent to CHP described above, but also to a 'bottoming cycle' where process heat rejected is used to generate mechanical or electrical power (Figure 4.3). So-called total energy systems describe industrial plants supplied with all energy by one fuel, broadened to describe CHP, cogeneration and waste heat recovery, implying optimum energy efficiency. In practice, however, achieving the perfect balance between heat and power to supply to properly and economically match the

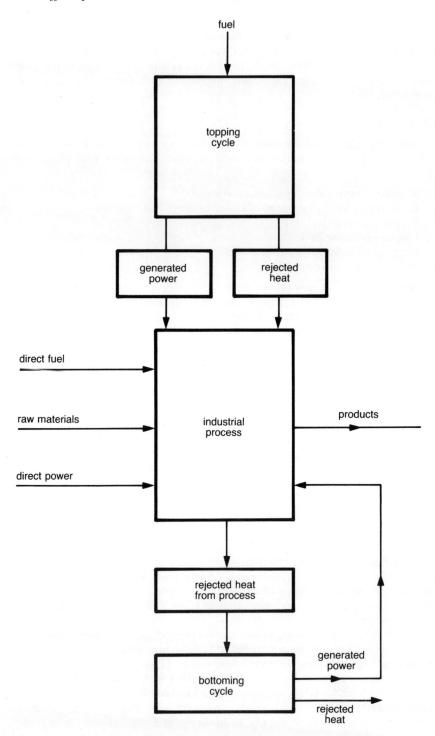

demand for each is difficult without using some supplementary heat or power source, demand control or all three. CHP ranges from a few kW in size to hundreds of MW, and capital investments from several thousand to tens of millions of dollars.

The thermal efficiency of fossil fuel boilers may exceed 80%; CHP has efficiency of only 75% but is preferred to boilers because of energy quality and not quantity. However, equating one joule of high grade electricity or mechanical work, with one joule of low grade heat is misleading. For a given demand for electricity and heat, CHP can use less fuel than if the same electricity and heat were produced separately. Demand for heat and power varies rapidly with time of day, etc., so detailed analyses of CHP variations plus several options of plant are needed to choose an optimum CHP design in any particular case. Figure 4.4 illustrates typical relative efficiencies of fuel use by separate and combined generation to meet a range of electricity and heat, base load electricity and steam requirements being the most economic here.

On the other hand, generating all steam requirements and some electricity only (i.e. in proportion to the steam requirements) could be a solution elsewhere. To be economic, CHP needs:

(a) Substantial heat loads with high load factors
(b) Small, simple heat distribution systems
(c) Generation sized for baseload heat
(d) Generation run in parallel with the grid.

In particular, careful attention is needed in costing item (b).

4A.3.3 An example

The UK 1983 Energy Act provided a good example of a measure designed to stimulate competition and encourage CHP, allowing anyone to provide electricity as a main business, requiring utilities to publish tariffs for purchases from private generators, using normal utility transmission and distribution as common carriers. The Act required utilities to support economic CHP schemes, bringing innovation to electricity sector CHP approach. A CHP bureau was established for co-ordination and publication, senior managers designated in each utility to advise on:

(a) Micro-CHP, a few kWs, based on automotive gas engines, for commerce and housing blocks

Figure 4.3   *Industrial CHP (Source: WOOLACOTT, P.G. 'Industrial combined heat and power prospects and developments', UK Department of Industry, 1986. Reproduced by permission of HMSO)   Illustration on previous page*

94 *Efficiency, conservation and the environment*

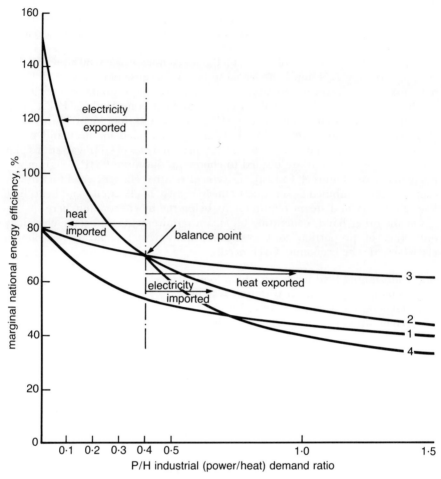

Figure 4.4 *Efficiency of generation of heat and power*
  1 separate generation
  2 combined generation (heat limited)
  3 combined generation (power limited)
  4 combined generation (P/H > 0.4 heat dumped)
  (Source: WOOLACOTT, P.G. 'Industrial combined heat and power prospects and developments', UK Department of Industry, 1986. Reproduced by permission of HMSO)

(*b*) Industrial CHP of 1 to 10 MW, using gas turbines, steam plant or even reciprocating engines
(*c*) Larger schemes, involving district heating for industry and the inner cities.

Such an Act only became workable after the privatisation of UK electricity.

Table 4.1 *Demand loss multiplier versus load factor (Source: 'Energy efficiency: optimisation of electric power distribution system losses', World Bank Energy Department Paper No. 6, May 1982. Reproduced by permission of the World Bank)*

Load factor	Loss factor	Demand loss multiplier
30	20·6	1·46
35	24·6	1·42
40	28·8	1·39
45	33·3	1·35
50	38·1	1·31
55	43·1	1·28
60	48·4	1·24

4A.4 Improving network efficiency

Extra high voltage transmission, say above 150 kV, requires quite definite minimum standards of design, construction and operation, thereby automatically incorporating only a very small electricity loss. World Bank publications on optimising other network losses, incorporating useful case studies, conclude that [5]:

(a) Loss reduction measures for networks provide extra capacity at less cost than do new generation and transmission
(b) Economic loss levels are lower than those traditionally accepted by utilities, thus losses should become a dominant fact, alongside reliability, safety, etc., in network design and operation
(c) Methodologies exist to determine economic loss levels of electricity apparatus.

The first problem in improving network losses is in determining if loss levels are within acceptable limits for the 1990s. Demand (kW) losses at peak are of most concern, requiring new investment at all network levels if this is sustained. Most utilities collect kW loss data and these losses must be forecast for the total electricity generated, estimating losses and peak demand over selected time periods, load factor, and assumed relationship between load factor and loss load factor. As a first step, system load factor and average and peak energy loss can be estimated using well-known expressions:

$$\text{Load factor (\%)} = \frac{\text{Energy generated (kWh)}}{\text{Peak demand (kW)} \times \text{hours}} \times 100$$

$$\text{Average energy loss} = \frac{\text{Energy loss}}{\text{Energy generated}} \times 100$$

Table 4.1 provides typical values of the demand loss multiplier (ratio of peak period to average loss) for various load factors. Thus, if a utility provides the following for a selected year (say):

Peak demand = 365 MW

Energy generated = 1,278,960 MWh

Energy loss = 217,423 MWh

$$\text{Load factor} = \frac{1{,}278{,}960}{365 \times 8760 \text{ hours}} \times 100 = 40\%$$

$$\text{Average energy loss} = \frac{217{,}423}{1{,}278{,}960} \times 100 = 17\%$$

then the approximate demand loss at peak may be computed from the multiplier of Table 4.1: 17% × 1.39 (multiplier) = 23.6%. Table 4.2 shows system loss levels, a total peak loss level of 12.0% being good.

4A.5 Improving utilisation

More efficient energy usages, and more investments in the right projects to achieve this [6] are needed in both domestic and industrial sectors, coupled with some fuel substitution. Blaming market failure for inefficient energy utilisation is still common. Individuals and firms optimise the use of electricity, assessing costs and benefits of alternative usage, as they do when assessing other goods and services. If markets were generally efficient this should mean making optimal use of electricity. Market deficiencies are: no futures markets in electricity, distorting externalities from other sectors, poor information flows, monopolies, inadequate property rights and irrational consumer behaviour. This theme is pursued later. Comparing [7] usage efficiency in the categories of residential space heating, residential water heating, residential refrigeration, lighting, commercial space heating and cooling, and industrial and commercial

Table 4.2 Demand losses (% of kW generated) (Source: 'Energy efficiency: optimisation of electric power distribution system losses', World Bank Energy Department Paper No. 6, May 1982. Reproduced by permission of the World Bank)

System component	Target level		Maximum to be tolerated	
	Within	Cumulative	Within	Cumulative
Step-up station	0·25%	0·25%	0·50%	0·50%
EHV transmission & station	0·50	0·75	1·00	1·50
HV transmission & station	1·25	2·00	2·50	4·00
Sub transmission	2·00	4·00	4·00	8·00
Distribution station	0·25	4·25	0·50	8·50
Distribution primary	3·00	7·25	5·00	13·50
Distribution transformer & secondary	1·00	8·25	2·00	15·50

motor drives, in the countries of (then) West Germany, Italy, Japan, Sweden, UK, and USA, shows important cross-national similarities, together with cognizance of recent significant greater usage efficiency. This is evidence of increasing electrification of loads, electricity accounting in the late 1980s for one-third of all energy in those countries. Potential for further improvement in each of the above six electricity usage categories are in making improvement generally in electricity usage, say by 10% to 20% by 2010; lighting by 30% to 50%, residential refrigerators by 15% to 30%, the other four categories by 5% to 15%, the smallest improvement being in motor drives. However, improvements are realisable only when governments and utilities both adopt the necessary policies, e.g. setting prices near marginal costs and using international prices for fuels.

Market imperfections often signal to governments to initiate electricity usage research, especially in developing countries where government co-ordination is needed to knit together small research efforts carried out by many firms and consumers for whom capital markets work badly, even when savings from improved utilisation efficiency exceed capital costs. Low-income consumers are unable to borrow for more efficient usage technologies, another case for possible government intervention. Government intervention must always be cost-effective, but never more so than when enforcing minimum utilisation efficiency standards. Setting such standards involves making trade-offs between consumers by class and by type. Optimally, marginal benefit to consumers rationally preferring more efficient appliances must just equal the marginal harm done to those who would prefer less efficient appliances. Efficiency can often be improved by making standards variable, e.g. nationally and/or locally, tied to major causes of variability in each case (e.g. climate, local capital costs, local fuel costs, local industrial outputs, etc.).

4A.6 Financing improvements

Concerning paying for improving energy efficiency, the financial groundrules of the 1990s mean utilities will willingly pay only for schemes with good, short-term financial pay backs to them, leaving longer-term economic pay back schemes to be pursued by government, government offering some incentives, preferably financial, to utilities if they wish such longer-term schemes to be pursued. Third party financing [8] of energy efficiency investments include leasing, instalment sales and shared savings; in common with all types there is a third party, i.e. other than the plant owner, installing the efficiency equipment at their own expense and assuming varying degrees of risk. Offering most potential is leasing, in which savings are guaranteed, in particular shared savings between parties concerned. Not only is initial funding transferred to investing parties from the plant owner, but so also is the investment risk. In typical shared savings arrangements, a service company conducts audits of particular facilities. If these offer scope for adequate utilisation savings, the plant owner is offered a contract under which the service company agrees to install the specified efficiency equipment to meet predicted savings at no initial owner expense. Normally service companies also agree to maintain and repair that equipment. In return, owners agree to share resultant savings at an agreed rate and over an agreed period (e.g. 50/50 over five years), measured from a baseline consumption determined by analysis of past electricity efficiency and consumption, amount and type of equipment, weather patterns and occupancy rates. Shared savings contracts include provision for ownership of installed equipment reverting to owners at contract end for either a fixed charge or a fair market value. Shared savings arrangements are performance contracts, customers only paying for what is delivered in actual electricity savings. If equipment does not perform as predicted, service companies stand to lose everything. Thus such contracts are a powerful motivator to a service company whose livelihood is wholly dependent on ensuring adequate savings in electricity usage. Some service and DM systems manufacturers offer third party financing, guaranteeing savings from improved efficiency, investment coming from outside but with payment on pre-arranged fixed bases, savings either guaranteed or at least equal to the fixed payments.

These financing techniques offer plant owners numerous advantages over conventional financing because:

(a) Owners do not need capital up-front, particularly important when capital is constrained.
(b) No initial capital is invested, and often no fixed payments contracted; finance is 'off balance sheet', attractive to institutions and industrial/commercial consumers not wanting efficiency investment to affect existing credit facilities or debt servicing
(c) Shared savings investors assume all risks that savings will in fact occur, particularly appealing to plant owners who are sceptical of efficiency technologies' ability to perform
(d) Facility owners are freed from establishing a competence to determine which technologies are correct for their needs.

Barriers to third party finance for energy usage efficiency improvements are:

(a) Lack of awareness of needs and benefits of usage efficiency improvement worldwide
(b) Need for complex contracts
(c) Difficulty of negotiating baseline consumption, often requiring accurate but scarce historical data
(d) Attraction of undesirable third parties to the favourable tax deductions for efficiency utilisation improvement investment
(e) Unattractiveness of third party finance for small firms because of overheads. (In the USA in 1984, 65% of third party companies had a minimum contract value of between US$50 000 and US$100 000).

4A.7 Commentary summary

Success in improving electricity efficiency depends much on government policy, which can vary worldwide from '*laissez faire*' to central planning. Such efficiency improvements have value in the 1990s especially in developing countries, for which unfortunately it is especially difficult to obtain capital for improvement projects, especially refurbishments. EHV grids hold little room for improvement, but generation, transmission and distribution in most cases find room for improvement, again mainly in developing countries. In utilisation there again seems room for improvement of efficiency, coupled with improving electricity markets directly or indirectly. Who should pay for efficiency improvement depends upon whether improvement is financially viable short-term, in which case utilities should pay. For long-term economic return governments must offer utilities incentives. Third party finance provides outside capital for schemes with a good chance of success.

4B CONSERVATION

4B.1 Introduction

Rival views on energy conservation conflict most where pressure groups obscure objectivity. Important conservation factors for the 1990s are: relative roles of conventional and renewable energy sources; feasibility and relative economics of old and new technologies; relevant energy pricing policies; abundance and continuity of fuel supplies; extent to which improving energy efficiency (section 4A) diminishes energy demand; and extent to which energy conservation should diminish demand. These factors also influence and are influenced by world and national economic growth, also world and national energy growth. Something, however, stands out; energy is likely to remain relatively expensive and could grow increasingly expensive, therefore it must always be used wisely. One way to do so is by employing all energy more efficiently (section 4A), encouraging all cost-effective measures to do so. The same is true for energy conservation, with which energy efficiency is closely allied. Major studies in the 1980s [9] highlight conservation's meaning and importance, the role of pricing and the continuous need for new conservation stimuli.

4B.2 National policy

4B.2.1 Energy shortage

Utilities' conservation policies are influenced by national, often international, opinions. Since the oil price rises, conservationists have been concerned lest continuing world growth, plus associated energy consumption, lead to fuel exhaustion (Chapter 1). These views have been successfully challenged, but most governments (including those of developing countries) in the 1990s will continue to call for reduced energy growth rates. Many programmes backed by many governments have been introduced to encourage energy producers and consumers to conserve, e.g. energy audits, information gathering and dispersal, financial incentives, demonstrations, energy usage labelling and energy standards. Governments have set up special bodies to encourage conservation, usually alongside encouraging increased energy efficiency, e.g. the UK Energy Efficiency Office in the 1980s. Even by 1982, many EEC countries were showing improvement in their energy elasticity (i.e. the amount of energy to sustain a 1% growth in GDP) of about 20%, since which year conservation has become lower key, even though EEC has called for a further 20% reduction in final energy demand by 1995.

4B.2.2 *Electricity elasticity*

Improvements in electricity elasticity are viewed as a measure of the success of increased energy conservation, plus improving energy efficiency, although electricity elasticity can be influenced by many and diverse things. The oil price rises did in many cases cause much conservation, because increased energy costs were passed through industry and commerce to GDP. Both energy and electricity elasticities, falling after 1973, slowed in the period 1982 to 1986, when oil prices fell. In the European Community in that period some energy elasticities actually increased; Netherlands by 4·3%, Belgium by 2·1% and Ireland by 13%. Both elasticities are expected to fall in the 1990s in developed but not in developing countries.

Important factors affecting directly electricity elasticities in the 1980s were as below:

(a) National structural changes, especially in developed countries, saw declining industrial production, for which electricity elasticities are high, and a dramatic rise in service sectors, for which these elasticities are low, e.g. steel and ship building declined while banking, insurance and tourism increased. Some developing countries are following the developed world in this matter in the 1990s.

(b) Product changes within manufacture led to concentration on specialised products for which the electricity elasticity is lower, e.g. high priced pharmaceuticals instead of basic chemicals, metal fabrication rather than gross metal hammering, electronics rather than electric motors; again, some developing countries are following suit in the 1990s.

(c) Changes in technology for industry, commerce and residences automatically reduced elasticities, e.g. automation, computer controls.

(d) Electricity took an increasing share of final energy consumption; up to the late 1980s this increased GDP because electricity is a 'prime' fuel by consumer standards, automatically commanding a higher price, reducing elasticities. For the EEC, electricity's share of the final energy rose from 11% to 17%, in the period 1973–1986, and it is still expected to rise in the 1990s.

(e) In some countries (e.g. the UK, Norway, the Netherlands and Indonesia) a major factor affecting electricity elasticities has been increasing production, sometimes quite sudden and large, of oil and gas, e.g. in the period 1975 to 1985 the UK electricity elasticity increased from near zero to 6.6% which is artificially high.

(f) Some deliberate attempts to specifically reduce electricity consumption have been encouraged, e.g. by improving electricity efficiency, just 'making do' with less electricity, and campaigns to 'switch off' electricity after usage.

Attempts must always be made in a particular country at a particular time, to understand various trends in energy and electricity elasticities, exploring what these mean, if such trends will continue and what will be the effects if they do. If reducing elasticities is the overriding national policy (e.g. by say 20% in the EEC) then a simple approach is just to accelerate existing trends of declining energy and electricity-intensive industries, e.g. steel, metals, shipbuilding, paper and pulp, etc., while increasing the imports of these from newly industrialised countries, e.g. Taiwan, Korea, Hong Kong, plus encouraging service industries. But the situation in the 1990s might well be more complex than this simple approach warrants.

4B.3 Electricity conservation

4B.3.1 *Meaning*

Improving energy efficiency can be a goal in its own right, (section 4A), but it will often involve energy conservation. Electricity conservation does not mean just looking for ways to discourage demand, or reduce quality of supply. It does mean:

(a) Using better, all national resources, including electricity, both from the national long-term economic and the short-term financial utility and consumer point of view
(b) Ensuring all kWh are produced, transmitted, distributed and used more effectively
(c) Using similar kWh to produce, deliver, and use higher standards and quality of electricity supply
(d) Using less (or more) electricity, if this is economically and/or financially worthwhile, e.g. giving improved resource usage.

In the 1990s marketing electricity can only be done properly through a consumer approach and this means a new emphasis for conservation, which in the past has been dominated from the supply side. Similarly to improving energy efficiency, demand management may well mean conservation. All three, demand management, energy efficiency, and conservation are interrelated, and it is pointless to try to completely separate them. One possible partial segregation is shown in Table 4.3.

4B.3.2 *Market conservation*

One extreme view of world commercial energy sources is that these are finite and that they will eventually disappear, in which case the conservation problem is to decide on the rate of the social time preference, i.e. whether to give most weight to present or to future consumption. This finite resources view makes the issue one of when the energy should be used up, rather than how, largely robbing commercial energy conservation of economic interest. In reality, options exist for substituting into renewable energy through electricity and higher-cost non-renewables, e.g. tar sands, nuclear. The main conservation practical issue for the 1990s is to achieve optimal balance between (a) exploitation in the market of existing energy sources, and (b) development of new sources. One method of quantifying the future price of primary energy is: (i) to assume future non-renewables will become exhausted and backstop technology must needs to be substituted; (ii) discount the backstop cost to a present value in the usual way, at long-term rates of interest; thus finding (iii) the difference between the extraction cost and the price corresponding to (ii), which is usually termed the user cost. On the basis of predictions of future backstop technology costs, and assumptions of the long-term interest rate, it is

Table 4.3 *Conservation and load management programme summary of costs and results (Source: Southern California Edison 1981 conservation and load management programme. Reproduced by permission of the Company)*

Line No.	Conservation and load management programme components	Annual cost ($)	Estimated results	
			Savings (kWh)	Demand reduction (−MW)
1	Non-residential conservation	8 863 200	1 264 787 400	327·8
2	Residential conservation	12 474 200	86 049 900	−
3	Solar	1 154 000	2 922 800	−
4	Public awareness	1 578 400	−	−
5	Advertising	879 600	−	−
6	Measurement	2 679 100	−	−
7	Management/administrative support	1 634 200	−	−
8	Non-residential load management	2 668 600	−	153·8
9	Cogeneration	885 000	1 092 800 000	158·0
10	Residential load management	4 363 400	1 368 500	134·7
11	Total C/LM	37 179 700	2 447 928 600	774·3
		1.52 US¢/kWh		48 US$/kW
	Other conservation and load management activities			
12	Streetlight conversion	6 702 000	27 356 000	−
13	Distribution circuit management	5 228 400	65 294 700	19·8
14	Conservation voltage regulation	2 150 200	1 727 369 000	91·0
15	Total C/LM	14 080 600	1 820 019 720	110·8
		7·74 US¢/kWh		127 US$/kWh

possible to calculate what the current price should be, given that marginal extraction costs are (say) US$2 per barrel of oil for low cost producers.

Table 4.4 shows how the correct current price depends on the expected long-term interest rates, given an assumed backstop technology cost of, say, US$100 per barrel equivalent in 50 years' time, e.g. oil from tar sands. The common view that current energy prices are always much less than they should be because of defective futures markets requires the assumption of a high backstop technology cost and low interest rates. If current prices are believed to be too low, then it would be beneficial to seriously hoard energy sources, which would reduce supply and impose upward pressure on price. Given that there will be scope soon for a futures market in energy, e.g. from real-time pricing (Chapter 6) in electricity, to accommodate different views on what is the correct conservation price, it is a research question whether the operation of the futures markets results in a current price which is too low.

Despite demonstrations of short payback periods and high returns on investments for electricity conservation projects, consumers and firms seem loath to undertake these. Also, firms often use stricter criteria for judging conservation projects than for normal investment in projects, e.g. demanding twice the return, which leads to a form of market failure. However, sound reasons do exist for firms according electricity-saving investments low priority, e.g. electricity is only a small proportion of total cost, or information on costs and benefits of such investments is expensive to acquire, or any conservation calculations are suspect because of difficulties with forecasting energy prices. Individual consumers also often have valid reasons for doubting conservation, e.g. it is known that not all costs are included in financial appraisals with convenience of home improvements being significant and the real return to householders being much less than calculations suggest; also it is misleading to look at one high expenditure item in isolation, it being necessary to take into account all alternatives; also response to national economic changes is lagged, with individual consumers and firms often subject to severe budget constraints. If banks and finance houses were willing to offer loans specifically for conservation this would overcome much of the present scepticism.

4B.4 Who pays?

4B.4.1 *Utility role*

Opinion polls carried out in the early 1980s by national gas, electricity and other consumers' councils found that three-quarters of the UK consumers turn first to their gas and electricity showrooms for conservation advice, half seek advice from local authorities and only smaller numbers from appliance shops. Most US utilities recognise their crucial role in conservation, advising consumers on using electricity efficiently. Some offer conservation to help control

Table 4.4 *Current price and long term interest rates (Source: SCOTT, A. 'Energy self sufficiency in the UK', Department of Energy, p. 127, (1984). Reproduced by permission of UK HMSO)*

Long term interest rate %	User cost US$	Extraction cost US$	Correct price US$
3	23	2	25
4	14	2	16
5	9	2	11

electricity bills, thus helping utilities to control their own costs, e.g. by reducing the need for more generation. Many utilities provide practical advice through energy audits and financial incentives to encourage residential, commercial and industrial consumers to save electricity. Much can be learned on who should pay for conservation by studying the attitudes of US utilities [10]. The three types of US utilities are: privately owned, commonly called 'investor-owned', publicly-owned; and co-operatives, the 250 investor-owned dominating and accounting for 80% of generation. There are about 1900 municipal utilities and 1100 co-operative utilities but accounting for only $8\frac{1}{2}$% of generation; state and federally owned utilities make up most of publicly owned capacity. Investor-owned utilities are the main generators and, being privately owned, bear fundamentally on any motivation to invest in conservation. They must always aim for short-term financially sound investment decisions since they borrow on the open market, compared to Treasury borrowing used by publicly owned utilities. Poor investment decisions directly affect a private utility's credit rating and thus the cost of its future capital. Many US utilities are investing large-scale in conservation because it is judged to be more cost-effective than building in new capacity.

Among US publicly owned utilities the six federal generating authorities are the largest producers, Tennessee Valley Authority (TVA) and the Bonneville Power Administration being the two giants. With the exception of the TVA, 60% of whose generating capacity is coal fired, these utilities sell from federally owned hydro plants, the electricity being sold to local municipal utilities and co-operatives, also directly to large industrial users, e.g. metal smelting industries in Tennessee and the Pacific Northwest. Over 1000 co-operative utilities were established primarily under the Rural Electrification Administration, set up in the 1930s to bring electricity to depressed rural areas, but their average size is small, accounting for only 3% of generation. Additionally, 70 privately owned combination utilities supply gas and electricity.

4B.4.2 *Residential conservation*

The USA Residential Conservation Service (RCS) was conceived following the 1973 oil price rise, to encourage conservation measures in homes, thereby reducing national energy growth. The service offered residences home energy audits, identifying cost-effective energy savings measures, costs and benefits of such measures; there is a tiny cost to consumers. The case for offering such audits was based on utilities already being in contact with all residential consumers and utilities possessing much more conservation expertise than consumers. Furthermore, the costs of providing audit services is but a fraction of total utility costs. Evaluating responses to audits shows that:

(a) Audit costs to consumers are negligible, to utilities small
(b) Characteristics of households requesting audits are significantly different from average, being more affluent, better educated and more energy intensive
(c) Response rates to audits varied considerably, utility motivation being critical, conservation-minded utilities market programmes aggressively accompanied by financial packages.

Yet audits on their own probably do not produce significant conservation. Researchers are finding that incremental savings of audited consumers compared to non-participant residential is about 2%–3% of pre-audit energy use, part of the savings being taken up in increased comfort. There is some evidence that differences in energy consumption between audited and non-audited homes increases in the second year after the audit, consumers waiting up to two years before installing recommended conservation schemes. However, despite modest levels of energy savings, home audits do motivate consumers to make conservation investments more promptly than they otherwise would, and such conservation investments are larger.

US utilities aggressively marketing audits joined to low interest financial packages have the largest participation rates for home audit and conservation programmes. Many consumers do not have ready cash to pay the sometimes high initial installation costs of conservation projects, and are unable or unwilling to use conventional bank financing for this purpose. However, substantial numbers of consumers will take out low interest utility loans for conservation. Three quarters of US loan participants would not have installed any conservation equipment without low interest loans, these consumers being among the most energy wasteful houses.

4B.5 Future outlook

During the late 1980s utilities developed approaches to manage demand (Chapter 3). In many cases conservation, with or without DM, served alongside new plant programmes. However, new planning methodologies are needed for the 1990s, integrating conservation with demand forecasting and DM, improved energy efficiency and consumer research. As more utilities use conservation programmes integrated with DM and dynamic pricing, to manage, massage and influence future electricity demand, the need for improving data, analytical methods and models also grows and additional research and development is needed. Additionally, consumer response, participation in exploratory load research and adoption of changes in electricity usage need to be better known. Governments have a role in increasing national and energy information flows so that market forces can operate more efficiently. Firms need potential benefits from conservation brought to their attention before doing anything; energy conservation information search costs are high for utilities and firms alike. The problem is to determine the amount of resources to devote to providing such information, bearing in mind that regulation may in the 1990s be an efficient alternative to information.

4B.6 Commentary summary

It is important to apply economic efficiency to conservation. If efficiency is ignored and instead is adopted a theory of value assuming energy is the only scarce resource, then undue concentration on just saving energy and fuel has significant effects on national output, often adverse. The electricity price currently often does not reflect its marginal social cost, and also price is not used efficiently because of various market imperfections. While many arguments about conservation have superficial appeal, on closer examination, many appear unconvincing, leaving the main candidates for serious consideration, economic comparison with new plant, backup to DM and improved efficiency, environment control, improved information flows and load research.

There is justification for governmental intervention if electricity markets are seriously defective; the level and type of government intervention still being a research question, but there is a general consensus that the prime role of government is to ensure that energy price reflects social and economic opportunity costs reasonably well. When energy is properly traded this means most fuels find their world market price, and there are two reasons why this is appropriate as a backdrop even to electricity pricing with respect to the conservation of all resources:

(1) It works; consumers adjust their consumption patterns, including conservation, to an energy price change
(2) An economic pricing policy requires neither omniscience nor intervention in decisions that a multitude of individuals have to make in a host of different situations, including considering conservation, i.e. it does not have to be deliberately selective, or distort the economy.

This view is too simplistic for many conservationists. In any case it is impossible to separate out costs and benefits from interrelated sets of programmes for DM, improved efficiency, conservation and environmental maintenance and related intentions, which means that any such programmes must be appraised on a joint product project basis, using methods which can only approximately apportion cost and benefits to one motivation.

4C ENVIRONMENTAL MAINTENANCE

4C.1 Introduction

Environmental degradation for the 1990s is a matter of general concern for developed and developing countries alike. Evidence is increasing that environmental management [11], far from being a luxury, is an essential ingredient for maintaining natural resources bases, upon which nations depend for continued economic development. Developing nations must therefore find a path to growth differing markedly from that traversed by their predecessors and learn from the latter. Developed countries must modify their behaviour, curbing excessive use of resources and managing waste more efficiently for growth to continue. Traditionally, environmental problems were limited to the effects of urban and industrial waste disposal on local populations, but critical environmental issues now include global warming, threats to the ozone layer, tropical deforestation, trans-boundary movement of hazardous wastes, acid rain, soil erosion, desertification and siltation of dams, all relevant in some way to energy. But these items also call into question the applicability of present financial and economic justification of conservation projects, especially with respect to developing countries.

In developed and developing nations alike, governments are increasingly strengthening environmental institutions, introducing appropriate regulatory or legislative mechanisms, and turning to using true resource costs in all economic appraisals of environmental projects. Utilising the useful concept of 'depreciation of natural capital' allows environmental deterioration to be translated directly into economic terms. The World Bank estimates that, although Indonesia's GDP increased by 7% per year between 1970 and 1984, the true growth rate is 4% per year, if such depreciation is allowed for. Many believe that pervasive environmental problems need a completely new approach, e.g. integrating environmental maintenance into policy-making at all levels, supplementing traditional approaches with socio-political weightings of economic parameters in cost-benefit analysis. Recognition also exists that special attention is needed to designing incentives so that they induce environmentally sound practices from the start of a project.

4C.2 National and energy policy

Making environmental issues part of the very make-up of national sectors' planning as suggested above, needs implementing through many activities, the first step being studies of the national economy, identifying basic environmental issues, underlying causes and options for dealing with these. The next step is to delineate ways of addressing these to achieve a consistent approach to their solution for all national sectors. For most countries, the conceptual step is to accept the need to prepare environmental policies to be undertaken, mostly by governments, but also with broad social participation, and hopefully international support to developing countries. For such countries these policies must cover the widest spectrum of activities, providing frameworks for integrating environmental factors into actual national economic and sector development programmes. Already experience is finding that such plans can be effective in helping all decision-makers, including in utilities, to set priorities for environmental action. In addition, such plans help raise public awareness and strengthen policy formulation on many critical social and economic issues. For developed countries greater selectivity and in-depth treatment is required, e.g. energy sector and subsector studies focusing on multiple causes. Sometimes it makes sense to address issues regionally, e.g. for regional communities such as the European, especially true where a resource base covers several countries. A major challenge facing any country or region in the 1990s is to integrate successfully outcomes of environmental studies directly into its own national economic, sector, or utility planning, which should form the basis for setting up priorities for dialogue between consumers, utilities and government. The environment is only just beginning to be reflected into such planning at all levels, relatively few efforts yet being made to trace likely consequences of resource degradation following on from past sustained economic or sector growth, or to identify feasible economic policy measures to deal with environmental problems. But by 1990 well over one-third of the World Bank's loans contained significant environmental components.

Normal methodologies for investment appraisal are appropriate for assessing environmental projects, e.g. cost–benefit analysis, setting prices according to marginal costs, using least-cost criteria for choosing between alternative environmental projects all of which are satisfactory technically, economically and financially to the nation, the utility and the consumer. Many environmental improvements require little investment, the parallel being with energy conservation in this respect. However, using conventional methodologies presupposes all costs and benefits are quantifiable sufficiently and accurately enough to base decisions on them, the parallel again being energy conservation, except that environmental factors need more study than conservation before costs and benefits are all established with sufficient confidence. However, the position on this is much better than many sceptics claim.

The most successful policy instrument by which to judge environmental projects is still economics, involving: resource–planning, resource-cost savings with and without energy substitution, economic efficiency testing, cost-benefit analysis, least-cost solutions, net welfare maximisation, economic returns, shadow pricing inputs and outputs, and energy pricing, all used in plant assessment (Chapter 7). However, when all economic sums using this mixture turn out to be adverse, this does not rule out doing that particular environmental action altogether; such an emotive subject often is justified on purely political grounds, but it is still important to carry out the economic analysis first. Parallels to this are in the cases of rural electrification and 'polluter pays' policies. In connection with the latter, it is never easy to ensure compliance with policy, especially with dumping toxic wastes from the energy and the industrial sectors, many countries having sound policies on this, but with little compliance. More importantly, most national energy environmental policies are concerned with the law or public finance, seldom with capital investment and utility or consumer finance, because energy benefits have a long maturity time and environmental benefits seldom come back to energy.

4C.3 Electricity policy

4C.3.1 *Environmental effects*

The major environmental effects from electricity occur in generation [12]. All thermal electricity production has major undesirable effects of harmful emissions and wastes; during accidents these may be very large. Hydro-generators involve riparian rights issues, problems in amenity and displacement of habitats, human and animal. The greatest worldwide effects come from global warming due to excessive carbon dioxide emissions, and problems of nuclear wastes. Global warming results from build-up of atmospheric gases allowing passage of solar radiation but reflecting back to earth radiant energy emitted by the earth's surface. With all gases emitted from the earth there occurs a process of gradual breakdown and re-absorption by oceans and biosphere. With the greenhouse gases, the rate of emission is presently greater than the breakdown and re-absorption rate. The four main contributing gases are: carbon dioxide, chlorofluorocarbons (CFCs), nitrous oxide and methane. The contribution of methane is difficult to assess since it is much shorter-lived than the other three, decaying in 10–20 years, compared to 200 for carbon dioxide. Nitrous oxide is not believed to make much contribution to the greenhouse effect, but is a serious cause of acid rain, and CFCs are being replaced in the 1990s, owing to their ozone-depleting properties, leaving carbon dioxide as the single major worry. Currently 1–2 billion tonnes of carbon equivalent is discharged annually into the atmosphere over and above that absorbed by the oceans or the biosphere. The burning of fossil fuels dominates the unnatural sources of CO_2 emission, coal being the largest contributor with a coal to oil to gas ratio of approximately 5:4:3. The USA contribute 20% of CO_2 emissions, China 10% and the UK, sixth in the world league table, 3%. Fossil fuels burnt for electricity production contribute one-third of the UK's emission and road transport 19%, the latter showing very strong growth; over the next 15 years half of the predicted increase in CO_2 emissions is expected to be by transport.

4C.3.2 *Remedies*

Although there have been some improvements already in the thermal efficiency of fossil-fuel burning power stations (section 4A), a thermodynamic limit exists and improvements are unlikely to be sustained without substantial changes in technology, the most promising being for boilers, pressurised fluidised beds, atmospheric pressure circulating beds, but also integrated gasification combined-cycle plant, with or without the 'topping' cycle (Chapter 7). Of these the topping cycle of combined pressurised fluidised bed combustion and gasification makes the most significant contribution to carbon dioxide reduction. Substitution of natural gas or oil for coal in a cycle of the same efficiency reduces CO_2 emitted by 40% and 20%, respectively. However, gas-fired combined-cycle plants make significant contributions to reducing emissions likely to be only a short to medium–term solution if based on known amounts of available gas reserves. In the UK, CHP if used to supply 25% of UK homes and offices would result in a reduction of eight million tonnes of carbon annually. Other alternative fuel methods need further research. Methods of CO_2 removal examined (e.g. selective membrane separation or absorber stripper systems) would seriously affect the economics of fossil-fuel plants, probably doubling both capital and operating costs. There is also the problem of disposal of the removed carbon dioxide itself.

Concerning non-CO_2 emitting fuel sources, the main contenders are nuclear, wind, hydro, wave, geothermal and biomass. Major reductions have been achieved in CO_2 levels emission in recent years in France and Scotland owing to increased nuclear power, but the current consensus is that nuclear is uneconomic if all costs and benefits are included [13]. Any serious global commitment to reduce CO_2 emissions, and costs needed to achieve this, may display nuclear more favourably. While most global large-scale hydro opportunities have been studied, wind, wave, geothermal and biomass energy sources still all need much study. Wind is currently believed to be the most promising; wave power does not at present seem economic; geothermal technology is still at an early stage but its potentiality exists; biofuels are beneficial to the carbon dioxide cycle. Trees during initial growth absorb considerable amounts of carbon dioxide, regenerated when burnt. Fast growing trees, such as willows, harvested after four or five years for use as power station fuel, would constantly recycle carbon dioxide. As much as 4000 km^2 of forest are required for 1 GW of generation. Another, under-utilised energy source is domestic waste. Technology is available for waste incineration-producing electricity, without emission of dangerous gases, and burning waste is doubly beneficial since it prevents the emission of methane produced from landfill waste sites. In 1990 Switzerland burnt 80% of domestic waste, Sweden 50%, France 36%, the UK 8%. The 1992 International Conference in Rio de Janeiro demonstrated the difficulties of defining and implementing any specific environmental measures.

4C.3.3 *Utilities*

Utilities in developed, and an increasing number in developing countries, attend to acid rain and greenhouse effects. In fact, acid rain abatement is undergoing research everywhere reducing uncertainties on source–receptor relationships and measuring ecological effects. Technologies for clean coal and oil seem reliable if costly. It seems to be agreed worldwide, even by utilities in developing countries, that gradual reductions of CO_2 emissions are needed, plus alternative energy measures which will maintain tropical forest and the planting of more trees. For nuclear generation, global repositories could be continuously built for high level wastes to very stringent standards over many, many years, reprocessing to recover plutonium, high level wastes then being less likely to leak. Inter-disciplinary teams are studying human health risks from electricity transmission, distribution and utilisation. Common action for all utilities should be:

(a) Major programmes for consumer education concerning environmental issues which are due to electricity production, transmission, distribution and utilisation
(b) Establishing strong channels among the media, scientific and engineering communities, ensuring that all environmental judgements are based on the best available advice on science and technology, finance and economics
(c) Establishing agencies for information interchange between utilities within and outside electricity, also between utilities and consumers.

4C.4 Economics and who pays

4C.4.1 *Markets*

In the right maket, prices reflect social opportunity costs and benefits, people using resources bearing and being aware of the social costs of doing so, adjusting their behaviour accordingly, this being behind the 'polluter pays' principle. Such does not occur where uncontrolled access exists to resources and people do not pay full social costs, e.g. power stations and industrial emitters of greenhouse gases and toxic wastes using the atmosphere, land, and oceans, as a free waste sink. Limiting access in some cases, e.g. to the atmosphere, is difficult to enforce. In such cases as the above, markets fail to allocate resources appropriately, there being incentives to overuse some resources, having no means of generating the revenues necessary for environmental maintenance or of preventing depreciation of environmental capital. With such market failures, some regulatory body is needed, but the information to regulate often does not exist, and what is normally settled for in practice are changes which obviously improve the status quo. Acid rain, or fallout from nuclear, is often not local or even national, but international, and for greenhouse gases global. This adds great complexity to markets devising pricing, implementing and monitoring environmental strategies, implying the need for international pricing, agreements and monitoring as well as local ones.

4C.4.2 Remedies

A range of possible policy responses exists for environmental issues, none costless:

(a) Form international groups examining evidence and considering policy responses, e.g. the Rio Conference of 1992.
(b) Change fuels; modify final emissions by apparatus.
(c) Reduce emissions by direct abatement; reduce fossil fuels for electricity; reduce electricity demand; use alternative energy to electricity
(d) Restock polluted waters, forests, etc.

Some policy instruments needed to do these are:

(a) Emission charges and taxes
(b) Marketable pollution permits
(c) Regulatory bodies
(d) Subsidies
(e) Direct public investment
(f) Zoning
(g) Education and propaganda

For greenhouse gases the 'carbon' penalty of various fuels, as mentioned earlier, deserves more serious consideration, although it is difficult to know how much it should be. Also, pricing is neither the only nor necessarily the best way of dealing with environmental effects in all circumstances, e.g. pollution taxes raise revenues which might not in the end be used to remedy environmental damage.

Governments have in practice shown much reluctance in using pollution charges, preferring regulation, zoning and often subsidies. Most polluters prefer not to face taxes, and there are reasons why governments tend not to implement them, e.g. the issue of fairness, a reluctance to subject certain groups to the explicit, visible impact of taxes. Internationally, these issues are of major significance.

4C.4.3 Nuclear

Looking at electricity projections, coupled to nuclear expectations which could mitigate so much the greenhouse effect, although widely divergent views exist on likely global electricity demand, there is general agreement that electricity will continue to increase its share of global energy and also its absolute value. Analysts differ in improvements in DM, energy conservation, efficiency, etc.,

Table 4.5 *World energy demand and electricity supply (Source: UKAEA Memorandum 15 to House of Commons Energy Committee Inquiry on Energy Policy Implications of the Greenhouse Effect', UK HMSO, (1989), 192-i, pp 96–103. Reproduced by permission of UK HMSO)*

	1987	2020 modest		2020 low	
Primary energy TWh/y	12·0	19*		15	
of which electricity					
(primary equivalent)	3·5	7		7	
Electricity generation					
1000 TWh/y	10·2	20		20	
		Postulated		Postulated	
Nuclear share %	16·2	21	50	21	50
Nuclear capacity GWe	298·0	700	1650	700	1650
Carbon dioxide					
reduction bn tonnes pa	1·7	4·2	10	4·2	10
Reduction in global					
warming %	3	6	14	7	16

* Based on World Energy Conference/International Institute for Applied Systems Analysis/IAEA low scenarios.

which will offset this growing electricity demand. Optimistically a global primary energy supply of say 11 tWh per year by 2020 might suffice, the same as 1980, but it seems unlikely to be less than 15 tWh per year. (1 tWh equals 1 million kWh, see Table 4.5.) There is closer agreement that world electricity demand will double by 2020. On a trends-continued basis, 500 to 700 GWe of nuclear capacity might be installed within the West, compared with 340 GWe expected to be in place by 2000. A 700 GWe for global nuclear capacity in 2020 means 20% of total electricity, only reducing global warming by 6% due to the small contribution of electricity to total greenhouse gas emissions (Figure 4.5). Nuclear substitution to achieve 50% penetration globally by 2020 requires a further 900 GWe capacity, 1600 GWe total, constructing 60 GWe per annum nuclear post-2000, plus 30 GWe p.a. replacement for plants erected in the late 1970s and early 1980s. This total GWe p.a. would succeed, if electricity does double in demand as generally agreed by 2020, in holding greenhouse gas emission from electricity at 1989 levels, contributing to the reduction of total greenhouse emissions by amounts such as 15% depending on total energy growth (Table 4.5). If energy efficiency improvements, etc., limit demand growth, then nuclear power contribution would be proportionately greater. However, since energy use only accounts for 45% of global warming, an upper limit exists on what is achievable by fuel substitution or energy conservation. 60 GWe p.a. of nuclear plant installed from 2000, rising to 90 GWe p.a. by 2020, would represent an enormous increase on current planning for the late 1990s, requiring major financial effort and firm decisions to be made before the mid-1990s, nuclear taking at least five years to build, and such amounts being an upper limit of achievability.

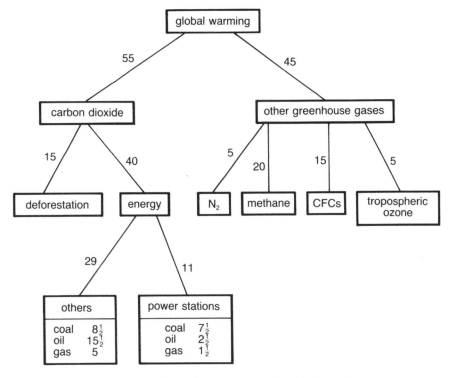

Figure 4.5 *Schematic mark-up of present man-made global warming*
Note:
All numbers given above are percentages of the total global warming effect of unnatural emissions integrated over two decades or so. These emissions enhance the warming due to the Greenhouse gases which occur naturally in the atmosphere. The global warming effect of a gas depends both on its concentration and how effective it is in trapping thermal radiation. It is worth noting that power stations account for about one-quarter of the energy contribution. Moreover, energy-related activities contribute perhaps 5% of global warming via gases other than carbon dioxide (methane, N_2O, ozone). (Source: JONES, P.M.S. 'Nuclear power, the greenhouse effect and social costs of energy', Surrey Energy Economics Centre Discussion paper SEEDS 51, May 1990. Reproduced by permission of the centre)

4C.5 Future outlook

4C.5.1 *Reducing emissions*

Global environmental problems need massive global collaboration, because individual governments acting alone would understandably hesitate in taking isolated domestic action, fearing that their national industries would lose some competitivity. However, international environmental collaboration is concentrated upon protecting endangered species and marine environments, with some atmospheric work, e.g. in the ozone layer and with sulphur, these being comparatively easy to monitor and reduce. Dealing with other gases presents greater difficulties. There are many problems related to popular schemes for equal percentage reduction of emissions:

(a) Such schemes are inefficient because with similar resources allocated differently, larger reductions are achievable to single or groups of emissions.
(b) It seems arbitrary that reductions are in proportion to late 1980s emissions, favouring late 1980s dirty technology users, penalising those already reducing emissions, leading to avoidance while awaiting for international agreements.
(c) It is not easy to find a fair base for permissible emission levels at all.
(d) Emissions proportionate to GDP favour the wealthy, proportionality to population favours the poor, both neglecting differences in technology, capital and other assets, taste and climate.
(e) The only scheme following from 'polluter pays' is some kind of market mechanism, taxes, charges or tradeable permits; the efficient solution to this problem of the 'common source' is payment to use that common resource, over-used if free.
(f) It is difficult to invent tariffs suitable for different pollution types, starting with CO_2 emission tax, because such emissions are global; but this does not necessarily mean that burdens should be borne equally; Bangladesh has more reason to be worried about flooding than Nepal, but the latter may be worse hit by increased solar radiation.
(g) Granting the right of taxation to some international organisation implies institutional problems concerning its legality, national sovereignty, acceptability treatment of non-participants plus global income redistribution.
(h) Effective, environmental taxes set reasonably high would generate considerable funds for any international environmental body, requiring careful consideration of its set-up, e.g. a new UN organisation or a major expansion of the UN Environmental Programme. To minimise risk a dual system is possible, some funds staying regional.

One important international aspect is the securing of a wide acceptance of, and participation in, measures adopted to avoid automatic relocation of heavy polluting industries in regions without pollution penalties, e.g. taxes. The rules of the General Agreement on Tariffs and Trade (GATT) could be used to ensure no party could obtain unfair benefits in trade by just allowing excessive pollution in its area. Developing countries are already badly hit by both environmental degradation and other problems implying lack of economic resources to carry out urgent environmental programmes and undertake R&D, never mind having polluters from abroad relocated in their territory. The above arguments show that, despite their economic plight, these countries should also apply the penalties, such as taxes, to pollution. However, aggregately, these countries should be amply compensated by redistribution of the proceeds of global taxes, from any international environmental body, e.g. to fund their huge environmental projects.

4C.5.2 Needs

Needed most urgently for the 1990s are:

(a) Better understanding of interrelationships between economic growth, energy and electricity demand and environmental effects
(b) Wide comprehensive ranges of inexpensive policy options for dealing immediately with energy and electricity caused environmental problems
(c) Institutions needed for using these policies
(d) *Technological options*

Environmental maintenance might produce its own electricity consumption, its own patterns of electricity consumption and both these effects also need understanding better.

Likely progress will be to build on established expertise, particularly in the methods governments, regulators and industry negotiate to achieve politically determined environmental objectives, control instruments and implementation techniques. Other themes will be relationships between regulation and technological change and the roles of the EEC and other international bodies. Two broad objectives seem likely. The first is to develop and refine analytical tools to provide better understanding of the consequences of different pollution control systems, in particular its industrial database, incorporating technical data relevant to plant emission characteristics and expanding the world power plant database to incorporate information on fuel gas desulphurisation and denitrification installations. Models associated with these databases need revising to allow simulation of environmental control. Further refinements must expand the range of choice available for emissions-reduction and permit the simulation of market based regulatory instruments. The second objective is to understand the political processes by which the environmental regulations are defined and to contribute to theories of regulation developed in the spheres of economics and political science.

4C.6 Commentary summary

4C.6.1 *Introduction*

World economic and energy growth will make for urgent demands for environmental maintenance in the 1990s. While total energy consumed by industrialised countries will decline due to DM, greater energy efficiency and conservation, it will greatly increase in developing countries. The total per capita demand for industrial and commercial energy is still, in the developing world, less than one tenth of that of the developed world, yet their populations are exploding and will soon be ten times larger, requiring massive energy inputs just to survive let alone grow. Overall, therefore, substantial energy growth is likely in the twenty-first century if world economic growth is to be sustained, mainly on account of developing countries. What effect DM, improved efficiency and conservation will ultimately have, and whether it is possible politically and financially for developing country governments to press for this, is difficult to see at present. It is even more difficult to see whether such governments can afford to respond to environmental maintenance requests unless financially assisted specifically to do so. World Bank loans directly for environmental measures will not be anywhere near enough to completely address issues from all energy projects, and, in any case, all such loans for environmental maintenance have to be serviced and ultimately paid back. Also, environmental maintenance itself tends to increase total energy because factors such as reduction in acid rain, CO_2 emissions, etc., decrease energy conversion efficiency, thus requiring more fuel. On the other hand, most simplistic approaches (e.g. predicting exponential world populations and energy growth) leave out self-correcting factors such as energy conservation, improved technology transfer, greater fuel substitution, economies of scale, and greater insistence on environmental maintenance from dedicated suppliers of finance and equipment.

4C.6.2 *Issues*

In the 1980s environmental maintenance began to be given a great deal of attention worldwide, e.g. problems of emissions into the air and water, damage to soil, damage to groundwater resources, micro-climate changes. Internationally less attention has been given to multi-country participation in programmes to address global warming, losses in bio-diversity, acid deposition, ozone depletion, etc. Most national and international environmental factors have important energy sector connotations which are presently being considerably worsened by lack of investment funds, especially in developing countries, for dealing directly with environmental maintenance. In all countries it is still much easier to get funds for new projects usually containing only minimal provision for the environment, than for direct environmental maintenance.

There are no reasons why methodologies used in normal investment analysis cannot be used when assessing environmental maintenance (e.g. cost-benefit analysis, rates of return and efficiency pricing), provided all costs and benefits are included adequately. Thus it is possible to find the overall least-cost environmental maintenance programme just as for any other least-cost project, i.e. that which has the maximum net present benefit to the economy or utility (whoever is doing the sums) over the lifetime of the project. Even when the net present benefit comes out low, some particular environmental actions might still be viable on grounds other than economic efficiency (e.g. for social justice, politics), parallels being in rural electricity, where economic efficiency calculations may well not include all positive socio-economic side effects. Also, presently, polluters may be paying strictly too much, i.e. as if there were no negative benefits from removing pollution.

It is seldom easy for national or international authorities to ensure compliance with policy instruments for environmental maintenance. Especially is this true for policies dealing with the toxic wastes from the energy and industrial sectors. In many countries there exist excellent policy instruments for specifying policies for this and similar matters but, unfortunately, there also exists little compliance. A more important allied point is that most national environmental policy instruments concerning energy deal with only either the legal framework or public finance, or both; seldom, if ever, are they concerned with capital investment finance, mainly because in the energy sector any benefits have long maturity times and the benefits from environmental improvements in any case seldom, if ever, come back to the sector agency. What applies at the national level also applies internationally, because most capital investment finance, especially when arranged for developing countries, is concerned almost entirely with loan details, not even touching on socio-economic costs and benefits; again this is especially true in the energy sector. This effect is compounded, especially but not exclusively in developing countries, by so many of them restructuring their economies towards short rather than long-term benefits, which does not suit most environmental maintenance projects, or components of projects.

References and further reading

1 WORLD BANK ENERGY DEPARTMENT (1982) 'Energy efficiency: optimisation of electric power distribution system losses' paper No. 6
2 MUNASINGHE, M. (1990), 'Energy analysis and policy' Chapter 2 (Butterworth-Heinemann)
3 See text books on turbines and thermodynamics
4 WOOLACOTT, R.G. (1981) 'Industrial combined heat and power prospects and developments' (UK Department of Energy)
5 Joint World Bank and United Nations Development Programme, 'Energy sector management assistance programme reports' issued in the 1980s; also, various reports of the World Bank Energy Department and various reports of UKODA and UKODI, see publications lists of such organisations
6 'Advisory Council on Energy Conservation, fifth report to the UK Secretary of State for Energy', Energy paper 52, UK HMSO, (1983); also UK Department of Energy (1983) 'Investment in energy use as an alternative to investment in energy supply' DEN/S/3; also Hilman, A. (1984) 'Conservation contribution to UK self sufficiency', *JEP Paper No. 13*, Heinemann
7 International Energy Agency (1989), Report 'Electricity end-use efficiency' also Hirst, E. (1990) Book review in *Energy Policy*; also reports in the World Bank's Energy Sector Assessment and Management Programme's reports, 1980 *et seq*; also Fisher, A.C. and Rothkopf, (1989) 'Market failure and energy policy', *Energy Policy*, August, pp. 397–406
8 Association for the Conservation of Energy, (1981), 'Report No. 3: third party finance', 9 Sherlock Mews, London, W1M 3RH, UK
9 Fifth report by the UK Advisory Council for Energy Conservation (1983), 'Energy paper No. 52, UK HMSO; also main findings of the European Commission's review of EEC Communities, Communication to the Council, (1988)
10 Association for the Conservation of Energy 'Lessons from America No. 2: US gas and electricity utilities and promotion of conservation', (1984); also, Pacific Gas and Electric Company 'Energy management and conservation activities' (1984); also, US Department of Energy, Washington DC, 'Residential conservation service evaluation report' (1984); also Synergic Resources Corporation, Pennsylvania, USA, 'The impact of the residential conservation service on natural gas and electric utilities', (1983); also, 'Energy audits as an investment: the residential conservation service', *Public Utilities Fortnightly*, April 1984 pp. 82–84; also Hirst, E. 'Evaluation of utility home energy audit programmes', (1984) Oak Ridge National Laboratory, Tennessee
11 WARFORD, J. and PARTON, Z. 'Evolution of the World Bank's environmental policy', *Finance and Development*, December 1989, Washington DC USA; also, Berrie, T.W. (1990) 'Can developing countries afford energy efficiency and conservation?', paper in the Working Group on Energy in Developing Countries, Meeting at Surrey University, Guildford, U.K.; also, World Bank (1989), 'World Bank support for the

environment: a progress report' Development Committee paper number 22 (World Bank)
12 PEARSON, P.J.G. 'Energy, environment and the market: energy failure and global issues': (1990), paper to *Conference on Energy, Environment and the Market* (Surrey University, UK)
13 UKAEA Memorandum to House of Commons Energy Committee Inquiry on Energy Policy: implications of the greenhouse effect', HMSO, (1989), 192i pp 96–103; also, Jones, P.M.S. (1990) 'Nuclear power and the greenhouse effect', paper SEEDS 51 Surrey Energy Economic Centre, University of Surrey, UK; also, Steiner, T. 'An international tax on pollution and natural resource depletion, *Energy Policy*, April 1990 pp. 121–124

Chapter 5
Public versus private

5A OWNERSHIP

5A.1 Introduction

Electricity generation is not necessarily a natural monopoly [1] because competition is possible, although few countries have been able to make it work. On the other hand, transmission and distribution are natural monopolies, e.g. it is cheaper to supply electricity by one network than by two or more parallel systems. However, consumers do not benefit automatically from having only one network, the few communities still serviced by competitive distribution do not charge more, utilities do not pass on even some of the lower costs from there being only a single network. Costs could also be reduced by using the same pole to carry rival electricity supplies. Therefore, even if transmission and distribution tend to be natural monopolies, there is no intrinsic reason for making them completely so. Furthermore, it is not logical to make network franchise agreements if retail competition is bound to fail. Many claim that electricity is a prime fuel, i.e. is an economic 'merit good' to society, therefore it should be supplied publicly, private utilities and all consumers not appreciating its true value to themselves or to the national economy needing, if anything, subsidisation from other sectors. However, electricity's special facilities are best dealt with under social costs and benefits in normal cost–benefit analysis. Scale economies are not a sufficient argument for having all publicly owned utilities, the latter, it is claimed, being the only ones to have sufficient funds to take advantage of large-size projects.

5A.2 Pros and cons

5A.2.1 *Against private ownership*

Five situations [2] traditionally make a strong case for private markets: (i) being supplied from public ownership; (ii) requiring complex, detailed 'heavy' regulations, i.e. situations where:

(1) There exist strong natural monopolies, e.g. electricity transmission and distribution, because monopolies could hold consumers to ransom
(2) Costs are decreasing naturally through scale economies, technological advances, improving production, transmission or distribution efficiency, because only a few large generators, or transmission, or distribution

utilities, could dominate the electricity markets, extorting high prices and giving poor standards
(3) Suppliers cannot be automatically charged properly for important aspects of their direct concern (e.g. pollution from generators, wirescapes from transmission and distribution networks), it traditionally never being easy to deal with such offenders who are privately owned, i.e. legally to enforce standards and set compensation (e.g. public opinion has hardened against privately owned nuclear)
(4) Customers are not easily chargeable for group supplies, e.g. public lighting
(5) Public concern exists about having higher standards than normal, these being unlikely to be fulfilled privately because of fear of lower utility profits but for which, in fact, most consumers are willing to pay, e.g. special lighting at road accident blackspots.

However, there are alternatives to having direct government intervention, e.g. consumer cooperatives or cogenerators with public charters, strictly regulated franchises, and binding service contracts with private utilities, enforceable at law.

5A.2.2 *Improving utilities*

By 1985, with 75% of world electricity still public, electricity sector efficiency was being described as ailing, economically, financially and institutionally. Many utilities were products of World War II and colonial rule. There were also moves towards privatisation to encourage new financial sources for electricity, especially in developing countries, hoping such investment and/or ownership would restore ailing utilities. In the late 1980s this trend towards private finance for electricity was encouraged, especially for developing countries, by governments restructuring their economies because of debt burdens and balance of payments problems. By 1990 there was a consensus worldwide of how to improve publicly owned utilities:

(a) Make much more efficient use, financially, economically and institutionally, of existing publicly owned facilities, especially generation.
(b) Ensure that this is done before adding any new major generation or transmission or distribution.
(c) Bring in private finance and ownership, but only as a part of a general restructuring of the national economy, the energy sector, and the electricity subsector, to improve economic and marketplace efficiency, debt service and financial flexibility.

5A.2.3 *Privatisation case*

From the above and other reasoning, privatisation can:

(a) Deal successfully with major items seriously hindered by being public, e.g. by removing most government interference of utilities, except perhaps for regulation, governments being notorious for overmanning, underpricing and tending to have poor institutions

(*b*) Help clarify related issues, e.g. utility accountability measurement, comparison of prices between energy subsectors
(*c*) Present government with a drastically different scenario of electricity from the traditional, for comparison with the traditional, and then provide a continual privatisation sector viewpoint.

One reason often given for privatisation is that better management exists in private sectors, but there are not enough first rate managers to go round all the world's utilities anyway, especially in developing countries and entrepreneurs are not at all used to the electricity sector. In any case, merely converting public into private electricity monopolies does not of itself confer benefits. Effective private fundings and ownership must produce structural changes on a dual route:

(*a*) Change the form of utilities ownership; but also
(*b*) Change the form of electricity markets.

Each route must be given equal weight and this may take considerable time and effort.

The main ways possible for changing utility funding and ownership, each needing careful examination and flexibility in adoption, are:

(*a*) Divesting directly into new hands
(*b*) Contracting out by leasing, financed from public sources but private firms then operating
(*c*) Establishing self management cooperatives, owned and managed by their own workforce
(*d*) Deregulating the electricity marketplaces under the claim of increasing their efficiency, all statutory barriers to competition being removed
(*e*) Forming group ownerships, several publicly or privately owned utilities and/or financial and industrial groups, owning all, or parts of the electricity sector and operating these either directly or through public or private agencies.

Many individual social, economic and political influences are important in choosing optimum privatisation formats. The general change worldwide towards privatisation since the mid-1980s has, in fact, been halting at best, and many dangers exist for the 1990s:

(*a*) As a point of principle, disposal of previously public assets to obtain a good sale can worry governments, voters and consumers.
(*b*) Underpricing these assets remains the main single practical worry, especially when sales are made to foreigners.
(*c*) Monopolies in private hands being another such worry, suggests it is better to retain public ownership and develop better profitability indices to ensure better accountability than in the past.
(*d*) Privatisations increase resistance to change.
(*e*) There remain many strong, continuing believers in traditional long-term economic public funding and ownership, rather than modern short-term, financial funding and ownership.

5A.3 Theory of the firm

5A.3.1 *Step by step*

To be strictly overall successful, privatisation must increase the net present value of electricity to the national economy and to consumers, not just to utility shareholders. Successful privatisation increases incentives at all levels of the national economy, as well as maximising utility profits. Treating the electricity sector as a large firm, or a close-knit number of small firms, reveals problems related to privatisation which are felt elsewhere than in one single utility:

(a) Unemployment may result from losing government control, although possibly temporary if the electricity sector expands.
(b) A sector in overall bad shape before privatisation may still not be commercially viable, especially in privatisation's early days. If it is virtually certain that long-term privatised electricity will be socially, economically and financially viable, then a distinct explicit subsidy must be given to all privatised utilities, and related organisations, e.g. fuel sources, to help with cash flow problems in the early privatised years, laying heavy obligation on the newly privatised utility to break-even by a given date.
(c) Disadvantages of swapping public for private monopolies can best be overcome by introducing competition whenever possible, starting off by removing statutory obligations to supply, any competition being in the interests of consumers and ultimately of shareholders.
(d) Replacing publicly-owned utilities' financial targets or annual rates of return on assets may prove difficult. In the absence of specially constituted accounting principles for privately owned natural monopolies, which will only emerge with time, an acknowledged [3] substitute for the early years of privatisation is for governments, or regulators, to allow annual increase in electricity prices in accordance with the formula:

$$\text{Price} = \text{Price}_{\text{year(n)}} \, (\text{RPI} - \text{X})$$

where RPI is the retail price index for the economy, X is a constant set at the discretion of government or regulator.

5A.3.2 *Priorities*

Priorities for privatising sectors will differ between countries, but electricity will usually be near the very top. However, if governments intend to privatise all publicly owned sectors, priorities may be by ranking sectors by:

(a) Total yield to national exchequer on privatisation
(b) Effect of their privatisation on balance of payments or the national debt, etc.
(c) Ability of inhouse senior management to prepare utilities for privatisation
(d) Degree of monopoly already existing, it being easier to privatise a utility with a monopoly than one without

(e) Ability to move long term towards 'pareto optimality', i.e. perfect competition, encouraged by entrepreneur type investment and dynamic pricing (Chapter 6).

Table 5.1 gives some idea of how priorities for privatisation can be set.

5A.3.3 Do s and don'ts

Experience with privatisation in many economies, especially developing, is still limited. In a survey recently carried out, of 37 developing countries [4] it was found that in all but two, the number of enterprises sold, leased or contracted out were fewer than 20, and the agencies usually had small asset values and employment; an equivalent picture from developed countries for the same period gave fewer than 30 agencies privatised for about 45 countries. Nevertheless, even such limited experience offers some clear privatisation lessons:

(a) First create a private sector policy environment that encourages efficiency.
(b) Resist offering privileges to public enterprises' buyers, because this runs counter to economic efficiency.
(c) Strengthen capital markets as a critical factor in selling public agencies, through offering shares to the public, because countries with weak or non-existent capital markets may have to sell public agencies through private placement, making it difficult to find buyers with sufficient capital.
(d) Be prepared for severe social effects, especially short-term, e.g. unemployment.
(e) Prepare well strategy, tactics and detailed programmes for privatisation, including government's role, concentrating on price, or asset value, or net benefits to the national economy.
(f) Give special attention to the dangers of undervaluing assets, most governments being inexperienced in managing privatisation programmes, requiring external experts to prepare and administer the programme for them.
(g) Build up transparent privatisation processes to sustain wise approaches in inevitable public debate, logging studies of alternative privatisation programmes, done with cost-benefit and financial analysis, proposed subsidies, implicit and explicit, truly revealed, in current and real terms, at present and in future, together with proposals about debt existing on the vesting date for privatisation.
(h) Improve agencies which will remain temporarily or permanently in public hands.

5A.4 National policy

In the early 1980s [5] the UK situation was as shown in Table 5.2 and Figure 5.1. The last column of the table shows percentage changes in manpower over two years. By this, British Steel, British Leyland, Rolls Royce, British Shipbuilders, plus British Airways and the bus companies had only small further savings to offer compared to the others, therefore being ripe for privatisation. Figure 5.1 demonstrates the type of government analysis needed

Table 5.1 *Main privatisation sales UK (Source: LITTLECHILD, S. 'Privatisation principles': paper to Conference on Energy Privatisation held at Surrey University, March 1988. Reproduced by permission of the University)*

Date	Company	Means and amount of sale		Gross (Net) proceeds £m
Jun 1977	BP	Offer	17%	564
Nov 1979	BP	Offer	5%	290 (283)
1979	ICL	Private sale	25%	37
Jun 1980	Fairey	Private sale		22
Jul 1980	Ferranti	Private sale	50%	55
Feb 1981	British Aerospace (1st)	Offer	50%	150 (46)
Jul 1981	British Sugar	Private sale	24%	44
Oct 1981	Cable & Wireless (1st)	Offer	49.4%	224 (182)
Feb 1982	Amersham Int.	Offer		71 (62)
Feb 1982	National Freight Co.	Mgt. buyout		54 (7)
Nov 1982	Britoil (1st)	Tender	51%	549 (536)
Feb 1983	Assoc. Brit. Ports (1st)	Offer	51.5%	22 (−37)
Mar 1983	Int. Aeradio (ex-BA)	Private sale		60
Mar 1983	BR Hotels	Private sale		45
Sep 1983	BP	Tender	7%	566 (557)
Dec 1983	Cable & Wireless (2nd)	Tender	22%	275 (370)
Apr 1984	Assoc. Brit. Ports (2nd)	Tender	48.5%	52 (51)
May 1984	Wytch Farm Oil (ex-BGC)	Private sale		80 + 135 +
Jun 1984	Enterprise Oil	Tender		392 (381)
Jul 1984	Sealink (ex-BR)	Private sale		66
Jul 1984	Jaguar	Offer		294 (288)
Aug 1984	Inmos	Private sale	76%	95
Nov 1984	British Telecom (1st)	Offer	50.2%	3916 (2436)
Apr 1985	British Aerospace (2nd)	Offer	59%	551 (345)
Aug 1985	Britoil (2nd)	Offer	58%	449 (434)
Dec 1985	Cable & Wireless	Offer	31%	933 (602)
May 1985 – Mar 1986	British Shipbuilders (Warship Yards) *	Private sale		117 + 40
Sep 1986	BA Helicopters	Private sales		13
Oct 1986	Trustee Savings Bank	Offer		1498
Nov 1986	British Gas	Offer		5434 (7556)
Feb 1987	British Airways	Offer		900 (871)
Jan 1987	BL/Rover (parts) **	Mgt buyouts		567 + 15 + 7
Jun 1987 – Apr 1987	Royal Ordnance	Private sale		190
May 1987	Rolls-Royce	Offer		1364 (1080)
Jul 1987	Brit. Airports Authority	Offer + tender		1225
Oct 1987	BP		31.7%	?
Aug 1986 – 1988	National Bus Co.	72? private sales and mgt buyouts		306 (est)

* Brooke Marine 0.1, Yarrow 34, Vosper Thorneycroft 18.5, Swan Hunter 5, Hall Russell ?, Vickers (incl. Cammell Laird) 60 + 40
** Unipart 30 + 15 + 7, Leyland Bus 4, BL Trucks 0, Istel (75%) 26. Beans Engineering (February 1988) 3.
Amount of Sale is 100% unless otherwise indicated. Where there are multiple issues, amounts may not sum to 100% where new shares have been issued during interim period.
Net Proceeds defined as gross proceeds, less old debt cancelled or plus new debt created, less costs of sale where known. Plus signs indicate further payments conditional on performance.
In addition to the sales listed, there have been Miscellaneous Sales totalling over £500m during 1979–84, plus revenue from North Sea Oil licences totalling over £200m during 1980–83, plus council house sales totalling several billions of pounds. (Source: mainly 'Privitisation, the facts', Price Waterhouse, 9 July 1987)

before embarking on privatisation of utilities, e.g. quadrants where utilities should then be locatable if privatised, not necessarily the same quadrants in which these would be placed before privatisation. Indeed, one principle government reason to privatise may be the necessity to shift a utility from an inferior to a better quadrant. Figure 5.1 thus classifies in a simple two-by-two matrix:

(a) Demand prospects good or bad, depending upon long-term trends
(b) Supply prospects, conducive or non-conducive to single or multiple or possibly competing, ownership depending upon technology.

Utilities in quadrant A were not expected to present monopoly problems as

Table 5.2 *Nationalised industries*, year 1981/82 (Source: BEESLEY, M. and S. LITTLECHILD 'Privatisation: principles, problems and priorities', Lloyds Bank Review, July 1983, Number 149. Reproduced by permission of Lloyds Bank PLC)*

Name	Turnover £m	Capital employed (CCA basis) £m	Workforce 000s	% change in workforce since 1979/80
Electricity Industry (1)	8 057	32 605	147	− 8
British Telecom	5 708	16 099	246	+ 2
British Gas	5 235	10 955	105	0
National Coal Board	4 727	5 891	279	− 5
British Steel	3 443	2 502	104	− 38
BL	3 072	1 521	83 (5)	− 31
British Rail	2 899 (2)	2 746	227	− 7
Post Office (3)	2 636	1 347	183	0
British Airways	2 241	1 338	43 (4)	− 24
Rolls-Royce	1 493	992	45	− 23
British Shipbuilders	1 026	655	67	− 18
S Scotland Electricity Bd.	716	2 817	13	− 5
National Bus Company	618	508	53	− 16
British Airports Authority	277	852	7	− 7
N Scotland Hydro Electric	270	1 981	4	− 3
Civil Aviation Authority	206	162	7	− 2
Scottish Transport Group	152	157	11	− 17
British Waterways Board	16	50	3	− 2
Total	42 792	83 178	1 627	

* These are the organizations classed as nationalized industries in the public enterprise division of the Treasury, as reflected in the White Paper Government Expenditure Plans, Cmnd 8789, with the addition of BL and Rolls-Royce.
(1) Including CEGB, Council and Area Boards. Figures for CEGB alone are £6 364m, £23 357m, 55 000, − 11%.
(2) Including government contract payments £810m.
(3) Including Giro and postal orders
(4) Reportedly 37 500 as at March 1983
(5) UK only; overseas approximately 22 000.

	Demand prospects	
	Good	Bad
Single	A Electricity distribution (Area Boards and Grid) Telecoms (local) Gas distribution Airports	B Rail Post (or possibly C?) Waterways
Multiple	C CEGB (excl Grid) Telecoms (excl local) Gas production Coal British Airways	D Steel BL Rolls-Royce Shipbuilding Buses

Supply prospects (row label for the table above)

Figure 5.1 *Classification of nationalised industries post privatisation (Source: BEESLEY, M. and S. LITTLECHILD 'Privatisation: principles, problems and priorities', Lloyds Bank Review, July 1983, Number 149. Reproduced by permission of Lloyds Bank PLC)*

multiple ownership was feasible; most were open to international competition, with recognised global excess capacity. Their customers would gain little from privatisation, only the taxpayers, yet private owners should, compared to government, be more:

(a) Willing and able to identify and rectify inefficiencies
(b) Willing to exploit new opportunities
(c) Able to reduce losses
(d) Able to free resources for other uses

Utilities in quadrant C had good long-term demand prospects, were very large and unaltered over considerable periods, having no monopoly difficulties because multiple ownership is viable, prime candidates for privatisation. In the UK, none were then organised as in quadrant C, all being single, as in quadrant A for the CEGB, less grid, and British Telecoms, quadrant B for the National Coal Board, and quadrant D for British Airways. Thus, privatising these involved:

(a) Recognising wholly or partly that they should really belong in quadrant C
(b) Securing benefits for consumers and employees without producing monopolies
(c) Needing careful attention for utility structures after privatisation, to secure maximum benefits for all parties involved

(d) Using main transmission as common carriers, preferably still publicly owned.

Utilities in quadrant A had good demand prospects, but local distribution has high sunk costs with no immediate technology challenge; it is also easily sustainable as local monopolies. Consumers were at risk, but franchises could be auctioned to private bidders, thereby transferring inherent natural monopolies to the seller, i.e. the government, benefiting taxpayers not consumers. Yet difficulties exist in specifying in advance to franchises what are proper supply standards. Furthermore, franchise contracts renegotiation costs are large, and it is difficult to sell franchises on the premises of the sustainable natural monopoly alone. Bidders want statutory monopoly privileges, but this does not mean nothing can be done in quadrant A:

(a) Restrictions on new entries can be lifted.
(b) Distribution can be divided into regionally independent units creating market pressures on suppliers.

Utilities in quadrant B had declining demand while supply favours having one utility. Substitutes for them all were emerging, and public ownership was applied to halt decline in services. Increasing subsidies pointed to gradual services withdrawal, but with profound socio-economic and political consequences.

5A.5 Electricity policy

Traditionally grid dispatchers load generators in increasing order of short-run marginal production costs, mainly fuel for thermal units, most hydro being fitted in as a replacement by thermal basis. However, if generators reduce output or shut down, their oncosts are automatically passed onto consumers and these can be large, e.g. nuclear and large steam startups and shut downs. With competitive utilities supplying common carriers passing on these oncosts is intolerable. Furthermore, most national utilities could make unregulated deals with large, monopoly, fuel suppliers and equipment manufacturers. Needing determination for electricity sectors is: (a) how much of the merit order savings are negated by such externalities: (b) what can take the place of merit order after privatisation; (c) who then ensures optimum operation, utilities or consumers; and (d) will the electricity markets then be cleared more efficiently at all times?

To transfer smoothly and efficiently electricity from public to private ownership it is likely to be impractical to:

(a) Make publicly owned utilities responsible for their own dissolution
(b) Change from public ownership to private ownership in one move, i.e. without, only as an intermediate step, first splitting large utilities into a number of smaller utilities, perhaps with private as well as public shareholders.
(c) Retain sufficient senior managers without having intermediate changeover steps, i.e. retain sufficient know-how in the process to enable privatised, smaller utilities to sensibly, smoothly and efficiently make the transition

128 *Public versus private*

(d) Avoid asset undervaluation without having intermediate change-over stages
(e) Deal with nuclear plant in a simple way, especially if connected with defence plutonium, when nuclear must generate a minimum number of kWh
(f) Experiment, poor privatisation being worse than none, perhaps especially for the grid acting as a common carrier or marketmaker (see later).

Figure 5.2 shows possible electricity organisation options.

5A.6 Utilities policy

5A.6.1 *Criticisms*

Electricity sectors, and especially generation, are usually naturally expandable and profitable worldwide, nevertheless they have not been immune from criticism, especially when in public hands, because of the following:

(a) Poor demand forecasting and slow response to changes in trend
(b) Consistent construction delays
(c) Wrong choice of generation size and type
(d) Expensive, excessive planning and operating plant margins to cater for uncertainty
(e) High manning levels yet low productivity
(f) Paying high fuel prices, holding fuel stocks
(g) Giving way too easily to government pressure, buying nationally, artificially increasing order rates, maintaining work loads of ancillary industries and manipulating public finance for their own ends.

5A.6.2 *Local ownership*

One remedy being put forward in the 1990s to overcome these faults, increase accountability and still leave generation in public hands is local power boards. Local board electricity ownership has been common in the past because it is sometimes half way between public and private ownership, with public and private shareholders, while retaining a form of public local accountability. There is some evidence that local power board utilities charge lower prices than others. But many seem to have greater capacity, spend more on construction, have higher operating costs, include fewer peak-related tariffs, relate price discrimination less closely to demand–supply conditions applicable to each group of users, favour business relative to residential consumers, and change prices too infrequently and in response to only large changes in economic determinants. They also adopt cost-reducing innovations less readily, maintain managers in office longer, and exhibit greater variation in annual rates of financial return. Overall, however, municipal or local power boards should be able to reduce the problems noted above with little incentive to engage in empire-building since they are more interested in securing cheaper and more reliable electricity, regardless of whether self-generated, purchased from other local utilities or independent producers, and in promoting new generation sources. Locally owned generators, however, have two further drawbacks: (i) profit incentives effectiveness depends on well functioning capital markets and it is possible for inefficient local utilities whose profits are low to be taken over;

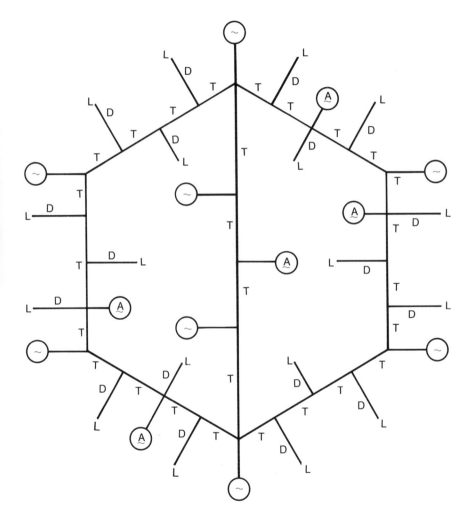

Figure 5.2 *Possible options in power system organisation*
⊖ generators, public or privately owned, unregulated
Ⓐ autogenerators, public or privately owned, unregulated
D distribution network, public or privately owned, regulated
T transmission network, public or privately owned, regulated
L large loads
Normally, 'market makers' to be the distributors and transmission network companies

and (ii) being semi-private might exacerbate monopoly power. Yet, the main alternative, franchising from a local base also has drawbacks, particularly in hampering capital market operation.

5A.6.3 *New entrants*

For generating utilities trying to get established there are problems of asymmetric costs, technology access and institutional barriers. Asymmetric costs occur when excess generation capacity occurs if most existing units have 20 to 30 years' economic life. However, in developing countries the demand is still growing rapidly and design improvements of conventional generators should make any new investment attractive in any case, even if only by a few years. Everywhere grids will be required to accommodate changing generation to demand balances, but transmission configurations for the early 1990s are presently often non-optimal, given demand management, conservation, dynamic pricing, and new fuel sources. By and large, however, globally present capacity is of a type which deters new generation, sometimes new transmission and distribution utilities. Also, although technology is changing in generation, potential entrants should not be disadvantaged since technical knowledge is widespread. Prospective changes in transmission and distribution technology are minimal. Concerning barriers to entry, statutory monopolies on generation have recently been lifted in many countries, but ready access to distribution as well as transmission will be necessary if such activities are to expand. Existing utilities have significant advantage over new entrants in this respect and for network wayleaves, and existing or potential sites; this could stifle new utilities.

Institutionally competition can be stimulated between new and existing utilities by:

(a) Having 'common carrier' obligations, similar to those in the USA from 1980, imposed on existing and new utilities, e.g. allowing non-adjacent utilities to transfer electricity between them across a third utility, or allowing new generators access to existing regional utilities beyond the confines of the utility on whose territory they are located

(b) Requiring regional generators to transmit electricity contracted for by large individual consumers, whether from new or existing independent generators or regional utilities; similarly for adjacent consumers, e.g. businesses on an industrial estate or households in a neighbourhood allowed to purchase electricity from other suppliers than the immediate generation utility

(c) Ultimately allowing non-adjacent consumers to form groups for electricity purchase on more favourable terms than those from the nearest generators, such groups acting as trade associations, or brokers specialising in knowledge of electricity markets

(d) Enabling connurbations, towns and villages to make contracts with various regional generators and local distribution networks, these to change hands in order to line up better with producers, at first the regional edges, thereby ensuring that more efficient generators, offering more attractive tariffs, expand at others' expense

(e) Prohibiting single national generators or at least subjecting them to restrictive practices legislation.

Generators in the 1990s will be tempted to play safe by signing short, medium, and long-term wholesale contracts, and selling electricity forward, laying-off risks by engaging in electricity futures markets, directly or through brokers (Chapter 6).

5A.7 Transmission specialism

5A.7.1 *Introduction*

Transmission in privatised electricity sectors takes on various new specialisms. Various types of ownership are [6] possible, e.g. by generation utilities, distribution utilities or both, or they could remain a separate utility. To arrive at what is best an examination of good specialism is necessary.

5A.7.2 *Common carrier grid*

This means distributors contracting directly with generators, the grid acting as an independent transmission medium, charging accordingly. Tariffs are then adaptable for transmission time-of-day with increased levies for unforecast transmission capability needed due to demand surges. As distributors thereby publicly answer directly or indirectly for supply security and reliability, grid utilities must also be legally obliged to provide not only adequate capacity to always clear the electricity markets efficiently at all times, but also a back-up margin to deal with unforeseen requirements. Ensuring that this margin is maintained, and preventing regular exploitation from surplus revenue requires heavy regulation (see later). US experience shows satisfactory long-term contracts between generators and distributors are difficult, jeopardising consumer supply (e.g. in the crippling New York blackout of 1977), perhaps especially in contracts taken out relieving strict obligations to supply in favour of reduced tariffs. Long-term contracts are calculated using long-term forecasts of costs, these often varying significantly from actual costs as variations in fuel price, weather, interest rates, etc., are not predictable. Break clauses are often introduced destroying guaranteed generation, distributors contracting for perhaps 50% more generating capability than required, increasing consumer prices. In the public interest all such safety margins should perhaps be decided upon by the regulator, but this kills competition and the advantages of privatisation are lost.

5A.7.3 *Marketmaker grid*

Autonomous grid utilities could buy and sell electricity from generators, and sell to distributors, but they must be obliged to supply distributors with predicted capacity at all times but, from above, without predetermined overcapacity plant margin. To maximise utility commercial viability of operation the purchasing policy from generators could work using merit order. The need for dynamic pricing is illustrated by failure of long-term contracting to reflect the real price at any one time, rendering it uncompetitive and needing higher contractual prices to offset unforeseen fluctuations. This involves generators quoting prices for various times through the following day, say a 24 hour prediction broken down into periods of 15 minutes. In setting a figure, generators would include all factors such as operating and overhead costs, incorporating revenue recovery and the quality of supply surcharge of dynamic pricing. Placing the competition onus on generators requires they not only operate efficiently to minimise costs, but also allows them freedom to add a higher percentage of levy at certain times of day (e.g. peak demand) and also at other times to compete at a lower profit margin. The link between generators and distributors needs legislation to set maximum allowable grid profits, i.e. limiting its monopoly. Assuming inevitable operation as close to this limit as possible, attractions of a grid utility to investors depends on values set by the regulator. If, however, the grid is obliged to supply transmission facilities to whoever requires them, grid margins will be squeezed to discourage distributing utilities from contracting directly with generators, reverting to common carrier.

5A.7.4 *Purchasing agent grid*

Contractual pooling is designed to overcome long-term contracting problems and protect consumers. A grid utility could accept all contracts made between generators and distributors subject to feasibility, and dispatch them in terms of the cheapest offered price available. The pooling of all electricity contracted for would then have two major effects: distributors would never be charged more than prices agreed in their contracts; and because of the vagary of demand, electricity would always be dispatched more cheaply than estimated. How to distribute the money saved is a matter for debate. Generators would not know competing prices, subject to commercial secrecy, thus eliminating cartels and minimising quotes. When pools produce profits, generators must have quoted unnecessarily high prices, competition ensuring larger generators lowering their profits, producing more and making more profit. Past wholesale bulk supply tariff problems (Chapter 6) are avoided if initial contracts are scrutinised by governments or regulators, ensuring they do not jeopardise competition. The regulator would be responsible for contract enforcement. Short contracts are needed to allow new generators entering the market, also keeping prices down.

5A.7.5 *Distributor-owned grid*

Each distributor can own equal grid shares with amalgamation by a holding company which has strictly limited operational control, thereby maintaining grid independence and limitation, dissuading operational cartels between distributors to exclude others or force prices higher. Implementation can be based as on market-maker, common-carrier, or purchasing-agent grids but, because owned by distributors, security responsibility lies solely in the latter's hands, practically advantageous because distributors are generally better informed about consumption trends, and can realise expansion of and modifications to the grid in time. Although this gives distributors power to buy electricity from the generators which can provide the cheapest electricity at any particular time, introducing competition and reducing prices to the consumer, it also points up the contracting process between the distributors and generators. Distributors are also allowed to import from outside the sector altogether. One disadvantage is the necessary restructuring of most 1990 grid control management structures. Operating in a publicly owned sector, the grid and generators form an integrated system managed together to prevent blackouts and to maintain the lowest production costs. However, with the above ownership patterns, especially with dynamic pricing, generators state their prices 24 hours in advance in 15 or 30 minute periods throughout the day, and the traditional control algorithms and procedures need little alteration.

5A.7.6 *Generator-owned grid*

This has substantial flexibility advantages for response to changing circumstances through dynamic control of demand–supply balances having the capability to switch generators in and out, avoiding drops in frequency and consequential possible poor electricity quality. This structure has performed well in the past (e.g. in France and the UK) in guaranteeing safe and secure supplies, but substantial disadvantages exist in implied lack of competition in generation, and the prevention of entry to potential competitors. To avoid the need for distribution utilities to build their own power stations, or to purchase supplies from independent cogenerators or autoproducers, single public grid generator-owned systems would need to be split up upon privatisation into a number of separate generation utilities in direct competition with each other. This ensures competitive pricing, a key privatisation purpose. The grid could be either controlled day-to-day by the largest generator or by a holding company jointly owned by all generators. Both have disadvantages in stifling competition, presenting barriers to new competitors and likely cartel formulation.

5A.8 Economics

5A.8.1 *Long-term contracts*

Before discussing the medium-term economic system involving private utilities (i.e. pooling), it is important to explore further the long-term electricity contracts mentioned earlier. Such contracts, under privatisation having no details publicly available, will be drawn up by distributors and generators under government or regulator rules, representing efforts to minimise risks, fix prices and give firm financial foundations to all utilities, especially just after privatisation. Electricity buyers and sellers will agree contracts covering most demand and all plant; although contracts will not be specifically related to plant, plant schedules will underpin them. A distributor of peak demand, say 5 000 MW, armed with numerous contracts, then has reasonable ideas about final electricity costs from main generators; hence main inputs to financial models are known, because contracts will be negotiated around assumptions about probable patterns of prices to be paid from the grid pool over the years. Contracts will usually be one-way options, capping the price paid for some electricity taken by distributors. In two-way options they will not gain full benefit from supplies which turn out to have a lower final price, and thus such contracts may be cheaper. The options are described in Table 5.3. The contract itself will have an initial price, after which other mechanisms such as pooling and settlements (see later) will take over, such contracts being between suppliers and distributors, i.e. nothing to do with the grid. The crude contracts likely to be established after any privatisation may be developed into more

Table 5.3 *Option types (Source: HANSON, P. 'The missing links', Electric Power Engineer, March 1990. Reproduced by permission of the journal)*

(A) One-way options
- If fixed rate [1] pool price [2] calls option up to maximum take
- G pays to D difference between pool price and fixed rate for these kWh
- D pays to G an up-front fee equal to projected value of rebates over contract life

(B) Two-way options
- Minimum take contract to reflect G's fuel obligations
- G gets pool price when fixed rate = pool price
- G gets fixed rate when fixed rate ≠ pool price

[1] Fixed rate = exercise price
[2] G-generator, D-distributor

complicated models related dynamically to load, plant and generating unit, buying electricity forward and setting up an electricity futures market. Nuclear electricity contracts, if nuclear is still 'special', may have a different form, e.g. relating to kWh actually produced. Although considerable stress is likely to be placed on having a computerised generator, grid and distributor settlement system running from the start of any privatisation, the first system used will never be final. Financial forecasts following privatisation will be based on initial long-term contracts and the prices agreed. A post-privatisation trial period for generation pooling will not actually help distributors much, and must have considerable risk because no public, comprehensive record at all will exist of plant availability which is determined by commercial decisions of newly privatised generators. Unless regulators take effective steps to stop cheating, a systematic long-term influence on electricity prices may be manipulation of outages, strategic mothballing of plant, etc., by generators.

5A.8.2 *Pooling and settlement*

Following the contracts approach above, and running at first in parallel with it after privatisation, will be a pooling and settlement system. Initially grid pooling and settlements will need two distinct features for the grid utility:

(*a*) Some merit-order dispatch system(s), possibly inherited
(*b*) A settlement system, providing a closing mechanism.

Generators will offer prices daily. Computer optimisations will initially be calculated assuming infinite transmission, the result being a system marginal price (SMP) calculated for each (half) hour, the market clearing price in commodity market terms. The SMP will be translated into a pool purchase price (PPP) by a weighted addition of the SM and a price of unserved energy, or value of lost load (VOLL). The weighting factor used will be based on the loss of load probability (LOLP), which itself will be derived from a daily computer assessment of relative plant and demand data for each (half) hour. LOLP runs may be performed on adjusted, declared plant availability to avoid commercial cheating. Thus:

$$PPP = SMP \times (1 - LOLP) + (VOLL \times LOLP)$$

where PPP is what generators actually get paid. Whereas SMP may be say US 6¢/kWh across the demand plateau of a winter day, VOLL is then more likely to be set by the regulator at say US 4¢/kWh.

Items added to the PPP to reflect system conditions are payment for reserve and ancillaries, adjustment for errors and misforecasts, etc. The resultant total will be then pool selling price (PSP), which distributors must pay for 'spot' market electricity each (half) hour. One school of thought envisages that the final computer system developed by the grid utility will be at least partly accessible to suppliers, grid and distributors, although this will be commercially difficult. Dropping distribution utilities' supply obligations would be ambitious as it will rely on market mechanisms and the operation of contracts to provide the required cost and price messages. Its success will depend crucially on determination or acceptance of long-term views of LOLP influences on pool prices. In this way, an implicit, explicit per annum price of capacity per kW will emerge, which in turn will lead to plant being built if generators consider the risk worthwhile, and if the right messages get through and in good time. The grid utility would need to outline initially its development plan for major consumers as a contribution to this process of synthesising long-term messages out of short-term effects. Further grid utility use of system charges will answer a claim to give additional spurs to generation construction, e.g. by getting new plant actually built in the right place to keep down ancillary costs.

5A.8.3 *Research needed*

Because there is no worldwide experience of the privatisation process, research is needed in the 1990s on:

(a) Economic benefits and costs of centralised versus decentralised control of electricity sectors
(b) Impact of privatisation on value of existing utilities
(c) Alternative privatisation formats, e.g. total privatisation including transmission and distribution, privatisation of only new plants, privatisation as such existing geographically-based utilities
(d) Likely price rises and falls, short, medium and long-term after privatisation
(e) Effect on investment of deregulated markets on utilities
(f) How generators should cater for additional entries from competitors when doing their own planning
(g) Income redistribution impacts of privatisation
(h) Best structure of electricity sectors in the transition to privatisation
(i) Legal issues
(j) Response of decentralised electricity sector to sudden changes
(k) How to measure marketplace efficiency
(l) How the grid should work
(m) Best regulatory systems.

5A.9 Commentary summary

5A.9.1 Lessons

In 1980 privatising [7] for economic change was barely acknowledged. For the 1990s the concept enjoys full recognition worldwide and governments embarking on structural economic adjustment increasingly see privatisation as an integral element. Their interventionist policies of the 1960s/1970s created many publicly owned utilities, now widely regarded as stumbling blocks to regaining growth. In the late 1980s the privatisation record was mixed. Even when successful, no blueprint emerges, arguing for a case-by-case approach rather than a model, yet lessons have been learned. The most commonly used privatisation methods have involved public shares offering, private shares sales, governments or utility assets sales, reorganisation into smaller utilities or a holding company plus several subsidiaries, and management and/or employee buyouts. The most popular has been private sale of shares and assets to single buyers, as in 530 recorded privatisation transactions in 90 countries. Many governments tried to privatise in national and sectoral economic environments, with domestic and international competition allowed to lead to efficient production, hence improving growth prospects. But the outlook for privatisation with high, uncertain inflation is poor; private investment then tends to remain dormant and prices stop sending signals that improve resources allocation. Arguments for privatisation often start entirely financial, e.g. shrinking budgets, mobilising finance managers, improving utility efficiency, frequently referring to economic benefits and a greater role for the private sector, and improving savings and investment allocations, hence growth performance. But the full financial–economic consequences cannot be measured or quantified. Moreover, many nations already short of cash, cannot stand the burdensome short to medium-term transaction costs involving financial restructuring or rehabilitation of utilities, redundancy and severance payments, restructuring or transfer of debt to government and/or the private sector, cost of advisory services, and the sheer time needed by busy government executives. In practice, therefore, political constraints frequently obstruct implementation of economic and financial policies designed to improve, through increased competition and more efficient performance, sectors in which privatisation is being sought. Again, privatisation has many enemies, posing a formidable challenge, the most organised and effective often being from unionised labour.

5A.9.2 Independent generation

Prospects for independent generators [8] can best be judged by:

(a) Latest technologies point downwards in scale, price and optimum size, away from 2000 MW towards 200–500 MW power stations; smaller units of the type available off the shelf, installed quickly and better for clearing future electricity markets

(b) The new small stations can be located within load centres allowing greater possibilities for cogenerators, autogenerators, CHP/DH and waste heat generators, raising efficiency from 37% to 75% plus, pointing the way to local power boards incorporating generators, distributors, cogenerators, users and independents, regardless of the privatisation process; independent generators themselves can belong to banks, constructors, contractors, utilities, etc.

(c) Long-term contracts for retail electricity will tend to be fuel-supplier dominated, and not usually truly independent generators, but rather ones which must sell to distributors or generate in consortia

(d) The stumbling block to independent generators everywhere for some time are long-term contracts and take-or-pay percentages in the value and/or shape of contracts, because few will get finance for anything other than take-or-pay long-term contracts for all the capital costs, preventing fair competition between generators and incoming independents, with no proper competition. With many sectors at first having surplus base-load generation, getting take-or-pay contracts to satisfy bankers on mainly peak-load power will be difficult.

5A.9.3 *Developing countries*

Privatisation is achieving limited progress in developing countries [9], the central issues not being desirability but feasibility. Fears of outside dominance, undeveloped capital markets, inflexibility in public finances, workforce opposition and private sectors highly dependent on government for subsidies and contracts, are all common constraints. Many utilities also have little asset value but high liabilities, and many aid donors and recipient governments see privatisation as a later stage, i.e. after public sector restructuring. However, halting progress does not halt the perception of privatisation as an effective government policy option. It seems clear that donors and borrowers will press for further private sector involvement, and the budgetary constraints, among other factors, will make ownership transfer, and other private participation, a necessary feature of economic reform packages. Probably also, privatisation will always be part, not necessarily a dominant part, of public sector reform, considerable prior restructuring of utilities first being needed. Though less newsworthy, such restructuring, rather than rapid de-nationalisation, is likely to be the major feature of developing countries in the 1990s (Chapter 8).

5B REGULATION

5B.1 Introduction

Regulation here means enforcing and monitoring operational rules to meet defined objectives by an appointed, autonomous, agency accountable to government [10]. Governments in the 1990s regard electricity as one of the public goods for which they bear at least some responsibility, seeking to control policies and operations of utilities through regulations, decrees, decisions by ministers or regulatory bodies. Regulation takes many forms; besides controlling prices and market access, governments control utility borrowing and investment programmes, generally restricting powers of directors and managers. Regulation claims ensuring efficiency, supply security, environmental maintenance and minimum bureaucracy. Yet there exists no general successful model. What has emerged is worldwide disillusionment with ineffective regulation; regulations have often been incomplete, unformalised, contradictory or open to arbitrary interpretation due to lack of regulatory expertise and resources. Sometimes wholesale utilities grow too powerful for regulators to prevent consumer exploitation by monopolies, wholesalers being less exposed than retailers to regulation. Typically, governments limit the actual autonomy of poorly performing utilities, in turn requiring technical efficiency and financial strength; then provoking government intervention in operational matters and increasing pressure on public finances for operational subsidies and budgetary grants. Widespread unsatisfactory utility peformance is raising questions about effectiveness of traditional regulatory control arrangements, especially for natural monopolies. New regulatory forms are badly needed which encourage competition, experience showing both government-owned and heavily-regulated utilities not to be efficient. This in turn encourages governments towards deregulation, or light regulation, providing sufficient autonomy to plan and operate on a fair basis with all other utilities. Some regulation is needed to limit profits and ensure adequate, safe and reliable electricity for all willing to pay. Private investors need reasonably guaranteed returns on investment but also must accept reasonable risk. Utilities need autonomy to compete properly with autogenerators, etc., and vice versa. Regulation needs to monitor utility operating performance efficiency.

5B.2 Types

5B.2.1 *Heavy or light*

Heavy regulation has:

(a) Several tiers, e.g. local, state, regional, national
(b) Large staff
(c) Powers to send for persons and papers
(d) Large numbers of papers from utilities
(e) Formal hearings
(f) A staff of many experts
(g) Wide subject matter to deal with besides price, including utility development programmes, financing and operational plans

Heavy regulation in the past has usually started light, the latter seen to be insufficient, often when some crisis externality suddenly became dominant, e.g. action by monopoly fuels or capital suppliers, hyper-inflation, environmental group pressures or gross price abuses. Because light regulation is the type usually first applied after privatisation, experience has often shown a natural progression towards heavy.

Light regulation has:

(a) No tiers and is centrally dominated
(b) Small staff needs for both utilities and regulator
(c) Infrequent contact between regulator and utilities
(d) Low profile
(e) Interferes for only a gross reason, e.g. on price

5B.2.2 Examples

In the USA there is heavy regulation, utilities being subject to rigorous examination of pricing, operations, development and financing plans, ensuring optimisation to the national economy as well as the utility. Accounts are open to scrutiny by Public Utility Commissions (PUCs), preventing cross-subsidisation, one of the basic tenets of US monopoly-utility franchise being to operate a reliable and non-discriminatory service. This degree of operations transparency is seen as vital to maximising economic efficiency for the nation, consumers and the utility. Utilities are given franchises to act as natural-monopoly suppliers in an area, but in return they must provide reliable services, meeting high standards. PUCs protect public interest but also ensure that utilities can earn adequate return on capital and assets. Above all, utilities must be seen to operate in the public interest, people concerned being empowered to intervene and cross-examine at all regulatory hearings. PUCs examine in detail utility investment proposals and, through certificates of 'Convenience and Necessity', can accept or reject them. In 20 US states PUCs require utilities to submit investment proposals in sufficient details to show that all alternative projects, including conservation, have been fully considered, and that investment is the least-cost available option in all cases, probably with stated compromises. PUCs have power to disallow investments, or expenditures, considered imprudent, or which do not ultimately produce electricity or conservation, because all investments must be 'used and useful' if consumers are charged their costs, ensuring that utilities make sound investment and that bad management costs are borne by shareholders, not consumers. Perhaps inside, but definitely outside the USA, heavy regulation is seen as expensive, slow, and inefficiently restricting utility management, but this cannot be proven. In business, strong regulatory forces, market or legal, tend to go hand in hand with strong, efficient agencies and vice versa. Some US states have only light regulation, and in these cases some utilities have made some unwise, costly investments resulting in higher costs to consumers and reduced profits for utilities.

The only reason or likely valid reason for having light rather than heavy regulation is economic efficiency, objectives being to:

(a) Encourage economic efficiency
(b) Manage in an efficient way any risks associated with structural economic changes accompanying privatisation
(c) Not scare off competition

In the foreseeable future, consumers will not globally enjoy direct choice of electricity suppliers, although this will come later through brokers. In this absence of marketplace competition, light regulation is the least that is required to constrain marketplace behaviour. Because all utilities are crucially important to national economies, policy-makers determining utilities' futures must now set up mechanisms for managing risks accompanying change before any changes are effected. Failure to do this would have significant effects on the national economy should assumptions made not be realised. Unrestrained, any monopoly in dominant market position will maximise monopoly rents, a monopoly compensating for inefficiencies through its ability to extract excess profit, if its behaviour is not restrained, e.g. if insufficient incentive exists to operate at maximum efficiency because there is no competition. Utilities account for a significant percentage of householder's net expenditure and are critical components in industry intermediate input costs. There are significant energy input costs both in raw materials for manufacturing and goods and services consumed by households. For energy-intensive industries where electricity may represent 25% to 30% of costs, impact of electricity costs on their competitiveness could be considerable. Pricing imposed upon all consumers, unrestrained either by competition or regulation and designed to maximise monopoly rent, would be likely to have a major damaging impact on any national economy.

Requirements to release information to ensure transparency of operation and ability to monitor operation is a powerful inducement to behave responsibly in the marketplace and thus justify light rather than heavy regulation. In support of the case for light regulation, transparency imposes its own discipline in market behaviour, and encourages economic efficiency, since consumers or user groups can judge for themselves a utility's performance. If the utility is judged to be exploiting a dominant marketplace position and/or operating inefficiently, public pressure follows or recourse is made to the courts or government in order to insist on changes in utility behaviour. It is possible in some circumstances that the mere threat of transparency with light regulation is all that is required to discourage anti-competitive behaviour. Provided the threat of transparency is real, not just a vague indication by government that it might change the rules in the future to counter misbehaviour in the present, it is a powerful inducement to ensure good utility behaviour. However, the threat of either light regulation or transparency will be empty without the provision of sufficient information to enable groups to appraise whether or not the utility is behaving in such a manner that the imposition of heavy regulation is required to constrain its behaviour.

5B.3 National policy

National aims on regulation are:

(a) Enhancing competition and preventing discriminatory practices, e.g. against new market entrants
(b) Protecting consumers from market abuse, e.g. by cartels and monopolies
(c) Promoting shareholder interest and confidence by ensuring worthwhile returns on capital or assets
(d) Maintaining security and high quality of supply
(e) Supporting government externalities, e.g. improved energy efficiency, conservation and environmental maintenance.

National regulatory policy is the most important influence on a utility's performance; the way the sector is regulated directly affects competition, efficiency, prices and the standard or quality of service. Weaknesses of utilities often arise from regulatory policy rather than from organisation pattern, e.g. independent generators have barriers preventing competition if main generators can ensure independents remain economically unattractive. The many problems inherent in regulating newly privatised utilities call for careful analysis to find the exact role of the regulatory body. Many governments attempt to reduce regulation by 'ring fencing' monopoly areas, i.e. separating these areas out financially for regulatory purposes. This should achieve the aim of liberating the utilities from any existing public sector restrictions on privatisation, while regulating the core monopoly aspects. But problems will lie in identifying and fencing the utility's core activities, ensuring these are not used to cross-subsidise unregulated aspects.

Regulatory bodies are organisable in two ways:

(1) A single framework acting as a 'unified regulator' for all economic sectors;
(2) Separate bodies regulating sectors or subsectors individually.

Properly equipped single regulators, applying different rules to different sectors according to their function, can achieve consistency and co-ordination across the whole economy, making maximum use of their manpower and financial resources. Having many regulators means being exposed to risks of inconsistency, inefficiency and possibly mismanagement. For these reasons it is normal to want one single regulator for at least each sector (e.g. energy) providing its senior staff are familiar with all parts of that sector and are capable of recognising the varying regulatory roles required between say electricity, gas, coal, and oil.

Until recently many governments' control over utilities consisted only through specifying external financing limits. External finance is capital requirements minus internally generated funds, and external finance limits are the maximum external finance that the utility can obtain during a year. Most treasuries divide capital requirements into fixed assets plus working capital charges. Internally generated funds are current cost operating profits, plus interest, tax, dividends, etc., but also including other contributions, e.g. funds generated from the sale of assets. External finance limits often conflict with other government proposals (e.g. recommended financial targets over a period), resulting in increased consumers' costs and excess profits provided mainly by small users billed quarterly, rather than large consumers billed monthly, which is hardly equitable. In the UK in the period 1980/81 to 1982/83 the financial target overshoot was £478m, and in 1983/84 to 1984/85 over £400m. In fact, the electricity sector overshot its financial target six years in a row at that time and the excess profits gave a financial boost attractive to future shareholders when the industry was privatised shortly afterwards. Figures 5.3 and 5.4 show the domestic and industrial tariffs (without VAT), the latter for various annual load factors with maximum power rating of 500 kW and 0·95 power factor. It seems that, like many publicly-owned utilities, UK utilities became complacent because there was no threat of bankruptcy, take-overs or pressures from shareholders. However, similar problems also occurred in USA regulated private utilities, indicating the problem is due to reasons other than public ownership. The idea that regulated privatisation is necessary or sufficient for improving electricity sector performance is thus questionable, not forgetting the expense of stockmarket flotations.

European regulation is variable, but always to some degree present. Regulation in (then) West Germany, Holland and commonly elsewhere is performed by rate-of-return calculations where the regulator fixes the price according to a maximum permitted annual rate of return of utilities on capital employed, or turnover, or assets. By limiting the utilities' profits, prices are expected to be controlled but it provides little incentive for utilities to reduce costs because the price they can charge under rate-of-return would thereby be reduced. An alternative European method sometimes used is to base regulation on price, setting the allowed rate of increase below the inflation rate in an '*RPI-X*' formula, where *X* is set to ensure maximised efficiency and *RPI* is the retail price index. Production costs should then be minimised.

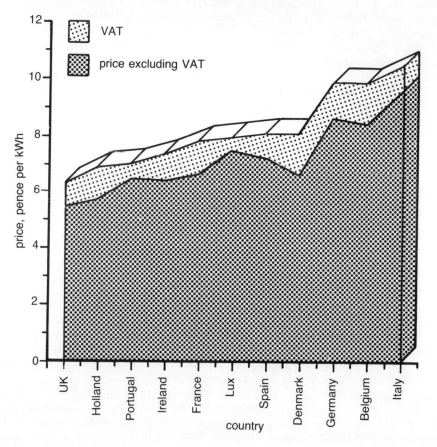

Figure 5.3 *Domestic electricity prices 1989 (Source: DUTTON, P. 'Operation of the new grid company', EE3 Group Project, Department of Electrical Engineering, Imperial College, London, 1989. Reproduced by permission of the College)*

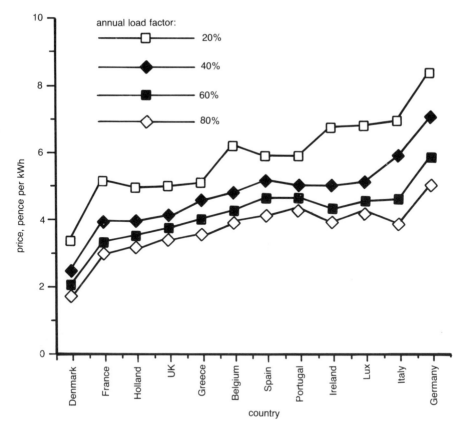

Figure 5.4 *Industrial tariffs 1989 (Source: DUTTON, P. 'Operation of the new grid company', EE3 Group Project, Department of Electrical Engineering, Imperial College, London, 1989. Reproduced by permission of the College)*

5B.4 Legal aspects

5B.4.1 *Transparency*

The importance of transparency in information has already been mentioned as being useful, at least at the early stages of privatisation; an information release level would then be legally requirable, giving sufficient transparency to allow for informed appraisal to be made of market behaviour and any utility's overall performance. This will, in turn, provide confidence to all that electricity is being supplied most efficiently. The nature and extent of the information legally required must recognise changes in the national economy and the energy sector, and the instinctive reaction of utilities to release as little information as possible to reinforce their dominant market position. What should be legal is the release of disaggregated, detailed, benchmarked and regular accounting, plus performance data, so that the monitoring agency charged with ensuring its release can analyse and publish knowing that the information level disclosed provides sufficient transparency to allow for meaningful and accurate appraisal. If not satisfied, the monitoring agency must be empowered to demand additional information and/or seek further amplification which it is entitled to publish or at least make known to interested parties. It is only this level of legal disclosure that will encourage economic efficiency, manage the risk of major structural economic changes and enhance competition. Anything less, e.g. merely the strictly legal provision of aggregate accounting and performance data, will not allow for meaningful and accurate appraisal. In addition, such information can be manipulated and adjusted or crucial matters disguised. In fact, disclosure of an inadequate level of legal information is more dangerous than no disclosure at all. 'Cosmetic' disclosure will disguise the true nature and intent of behaviour lying under the surface.

5B.4.2 *Establishing monitoring*

Important points when establishing monitoring are:

(a) Cost of compliance; additional costs involved in meeting legal information are justified because this information is needed for proper management anyway
(b) Cost of administration; administrative changes or costs are not major, insignificant in comparison to the national economic cost if risks are not managed and the new privatised structure fails to deliver benefits
(c) Asset value; care is needed to avoid a form of transparency which devalues utility assets privatised but with no way of quantifying this
(d) Excessive regulation; this can produce a legal proliferation of brokers and agents, impeding economic efficiency and discouraging new marketplace entrants; one way to avoid heavy regulation costs is to start with light regulation then establish an economic efficiency framework through the required degree of transparency.

5B.4.3 *Legal prerequisites*

The legal prerequisites of the monitor responsible for information release must be carefully defined.

(a) Accessibility: any individual or group representing users, consumers, potential users or competitors must be able to access the monitor for information, or complain if information is not provided.
(b) Independence: the monitor must be independent of political or industrial or commercial associations which limit obtaining and publishing information, or risk its credibility; those using the monitor's services must believe in its independence and impartiality absolutely, the monitor protected against capture by the sector being monitored.
(c) Powers: the monitor must have sufficient legal powers to get information released and published.
(d) Knowledge: the monitor requires a knowledge of the sector and the way it operates in order to access information and to identify and request more where it judges initial information is insufficient for required transparency.

5B.4.4 Location

The three monitoring prerequisites, accessibility, independence and power, require the monitor to have an independence backed by specific legislation. The fourth prerequisite, knowledge, can be provided by power to commission experts. Establishment costs of a truly independent monitor need not be significant and are far outweighed by long-term benefits in terms of meaningful encouragement of economic efficiency. However, monitoring conducted directly by government departments seriously compromises their independence, leading to conflicts between duties of monitoring and of advising government. If monitoring is placed within government, specific legislation is needed to ensure independence, specifying clearly the monitor's statutory powers and obligations, and thus avoiding confusion of public servants.

5B.5 Commentary summary

The precedents are not favourable when regulators [11] are allowed to examine critically utilities' costs of purchases and investments as well as prices. However, regulators should regularly require generation and distribution utilities to compare demand and supply-side options and to invest in options offering lowest cost to consumers. In the first years after privatisation, competition in generation is limited, thus if distributors can avoid purchasing generators' output by buying into DM, energy efficiency and conservation of resources, then, from the outset, generators will be competing at least with energy efficiency. Utilities are unlikely to seek least-cost investments for the benefit of all parties without initial stimuli from an independent regulator suggesting it. If necessary, it is more sensible for regulators to examine a utility's investment and financing plans at the planning process' beginning rather than after investments have been made, ensuring that utilities properly compare demand and supply, optimise resources, avoiding investments unwise to the economy, utility and consumers. Sometime regulators have their own career structures, separate from civil servants, leading to both continuity of management and stronger commitment to the organisation. Otherwise, regulatory staff are civil servants on two or three years secondment. Old-established regulators have drawbacks, tending to be over-legalistic, though not always bureaucratic as is claimed. It is widely believed that there is a strong correlation between purposeful regulators and efficient utilities, in the consumers' interests.

Strong regulators can encourage rather than stifle competition, competition occurring in an active 'heavy' regulatory climate, rather than in a light or passive one. With light or weak or passive regulators, utilities use monopoly powers to protect themselves from competition, with resultant economic inefficiency. Allegations that heavy regulators restrict abilities to manage efficiently are not well founded. Purposeful regulation of privatised utilities is needed, not to stifle initiative competition, but to ensure capital is used most efficiently to the benefit of consumers, utilities, shareholders, and the national economy; perhaps why there seems to be no intermediate point between light and heavy regulation.

5C FINANCING

5C.1 Introduction

Until the early 1990s, developed countries contained [12] various mixtures of publicly and privately funded and owned utilities, all making some contribution towards investment, their self-financing ratios varying considerably from below 20% to above 80%. Other funds were borrowed from central, regional or local government sources, under a variety of conditions, with lower interest rates and longer payback periods than commercial sources, which also provided some funds. Privately financed and owned utilities became common in the early 1990s in developed and developing countries alike, mainly to introduce private funding into utilities because public finance could not cope. It was believed that the private sector would introduce competition into utilities and improve marketplace efficiency. Private finance, however, has much more stringent terms than public, interest rates being higher and payback periods shorter.

5C.2 Financing types

5C.2.1 *Government guarantees*

Governments sometimes, especially for developing countries, support private credit by guaranteeing part or all monies being offered against a variety of circumstances and risks, e.g. default of lender to supply or default of creditor to repay, the safer of the commercial risks and the safer of the socio-political risks. Cofinancing is where funds from a principal donor—e.g. the World Bank or a finance house—are associated with funds from other sources outside the borrowing country, all funds financing parts of a utility programme. The principal donor actively promotes all funding. Much financing for electricity, in developing countries especially, is cofinancing with three partners:

(1) Official, including governments, government agencies (e.g. USAID), and multilateral institutions (e.g. the World Bank)
(2) Officially-supported (e.g. export credit institutions) socio-political finance for a particular country
(3) Private finance, e.g. commercial banks, finance houses, insurance companies, pension funds and purely private sources.

5C.2.2 *Foreign monies*

Commercial and official lending worldwide, but especially to developing countries, dropped dramatically in the late 1980s debt crisis, making the electricity worldwide financial volume inadequate. Foreign intervention became much encouraged, both as lending and as direct investment, the terms of which have striking differences:

(a) Borrower countries end up with liabilities to service loans that may or may not have brought economic growth; equity holders, on the other hand, get repaid only if their investments yield returns.

(b) Direct investment is tied to specific projects, not easing capital flight and aggravating balance-of-payments problems.
(c) Foreign investors put more than money into their projects, typically adding managerial skills, technical knowhow, manpower training and marketing connections, ingredients not obtained from commercial lenders or even always from official loans.
(d) Foreign investors develop local entrepreneurial capacity, stimulate local competition and engender upstream and downstream effects.

For all these reasons, direct investment has potential for high returns for both investors and utilities, mobilising foreign direct investment for good projects even in developing countries, despite balance-of-payments and other problems. Despite debt problems, investing in developing countries can be profitable, e.g. between 1980 and 1985 US foreign direct investment paid off at 11.7% in developed countries but 17% in developing countries, not counting benefits in expanded productive capacity that profitable projects generate. Decisions of investors to risk funds in developed rather than developing countries' electricity, however, often have less to do with markets than with business psychology, developed countries being seen as a safe proposition. Third World bidders for capital must overcome strong perceptions of political risks, a non-commercial barrier that decisively impedes intake of capital and technology needed to sustain growth.

5C.2.3 *Special role of MIGA*

The World Bank's Multi Investment Guarantee Agency (MIGA) was founded to break down socio-political barriers by reducing non-commercial risks. For member countries MIGA provides guarantees against restrictions on repatriating investment proceeds in convertible currency, also revolution and civil strife, conceivably even terrorism and sabotage; also insuring against government breach of contract and dangers of outright or indirectly creeping expropriation, confiscatory taxes, discriminatory delays in granting import/export licences, and similar licences, and similar factors. It extends insurance coverage to every equity investments type. However, insurance coverage is only the final product, MIGA issuing its guarantees in most cases after it performs a broker's role in bringing host countries' investment potentials together with foreign investors interested in taking a stake in productive opportunities.

5C.3 Financial engineering

Utility finance, investment and pricing policies are directly interdependent. If electricity demand–price elasticities are greater than zero, financing policy affects demand estimates and investment programmes since the costs of finance differ with source. Financial engineering starts with questioning:

(a) How is electricity development linked to economic growth?
(b) What are appropriate mixes of policies, institutional reforms, and investments to properly meet electricity demands?
(c) What are the most effective financing instruments?
(d) What will ensure that electricity utilities are properly funded?

(e) What role should government and the private sector play in their finance, investment operation and regulation?
(f) What effect will technological change have on investment and financial planning?
(g) What are the links between electricity development and the environment?

Asking these questions to best match the type of finance from those already described to the type of project constitutes financial engineering, only to be learned by experience. Crucial factors to bear in mind are:

(a) Project lead time
(b) Project completion time
(c) Amount of capital required
(d) Likely interest rates on loans
(e) Likely payback periods
(f) Likely management fees
(g) Loan repayment lead times

5C.4 Financial returns

All utilities should be revenue-earning entities and given the budgetary problems and capital markets of many countries, it is prudent that utilities finance a reasonable proportion of their expansion programmes from internal cash generation, say between 20% and 60%, depending on the external capital sources. Again, revenue should be raised principally by direct user charges. One of a utility's major objectives is to establish and maintain financial viability. In preparing a specific financing plan it should plan on precise criteria for minimum financial performance to assure continued solvency. Careful formulation of performance tests and of strict observance are in the utility's basic interests. There are three main criteria:

(1) Rate of return on investment, i.e. net operating income after taxes as a percentage of net fixed assets in operation plus adequate working capital
(2) Contribution to expansion, i.e. internally generated funds after debt service as a percentage of capital expenditure
(3) Operating ratio, i.e. operating cost as a percentage of revenues.

While the objective in all three cases is to provide the utility with adequate cash, the last two cases have shown disadvantages in their practical application: the contribution test has to use sliding averages of several years including estimates to overcome 'lumpy' capital expenditure and is almost impossible to test for adequacy. The operating ratio is limited because it cannot be applied to some common electricity systems e.g. hydro or mixed generation. It has therefore become common to express net revenue commitments in the form of a rate of return because it: (a) reflects generally accepted costing concepts in pricing; (b) relates net income to corresponding revenue producing assets; (c) mirrors concepts of regulation used globally; and (d) has proved a most convenient and objective test to apply. Acceptable rates are determined after the utility is satisfied that fixed assets are adequately valued and suitable arrangements exist allowing assets revaluation for inflation and to ensure that net income and depreciation charges are at the same level in constant terms. With chronic inflation, a rate of return or operating ratio test can only mean anything with periodic asset revaluations. Legislation may be required.

Definitions of reasonable returns have always aimed at providing sufficient internal cash generation. The determining factors for the return level are therefore existing capitalisation of the utility, expected growth rate, prospective investment patterns and availability of other capital. Emphasis on cash aspects of rate making is, in virtually all expanding utilities, consistent with the parallel objective of having rates include an adequate charge for the cost of utility investment capital. If an adequate return on capital generates more cash than the utility needs, excess funds are available to owners. But earnings by themselves rarely cover expansion costs, and are usually supplemented by loans or equity capital.

Debt limitation tests are important only when incurring new debt, objectives being to prevent the utility assuming debt service commitments endangering solvency, particularly important in ensuring utilities maintain sound financial conditions in view of their large and frequent needs for funds. The best limitation is a debt service test because: (a) it is directly suited to the defined objective; (b) it is based on figures clearly and objectively defined; and (c) it can normally be met with reasonable certainty because of the stable earning prospects of utilities. The test requires net revenues of a recent consecutive 12 month period (revenues less operating expenditure before depreciation and interest) to cover maximum debt service in any subsequent year by a specific margin, say 1.5 times. Actual revenues are usually nationally adjusted to take account of rate increases made effective after the beginning of the period and before the test is calculated. Due either to particular circumstances or to provisions in indentures, a number of other types of debt limitation tests are also used such as: debt/equity ratio; debt/assets ratio; earnings tests; coverage of interest by net income; debt incurrence defined as debt drawn down rather than debt contracted for; discounting future earnings increases, etc.

5C.5 Commentary summary

Financing requirements for electricity are likely to remain huge worldwide, as mentioned in many chapters, being made up from official and private sources. Official finance usually involves government directly or indirectly in making or guaranteeing loans, while private finance includes commercial banks, finance houses, stock markets, insurance companies, pension funds, etc. All have their advantages and disadvantages, and the art, learned by experience, of financial engineering is to find the optimum mixture of financial sources for a particular utility or development programme or project. Obviously the short, medium and long-term must all be catered for. Drawing up optimum financing plans is just as important as drawing up optimum development plans, from which they follow after making assumptions about construction costs, fuel costs, lead times, etc. For publicly-owned utilities, borrowing from government is common, often accompanied by raising money elsewhere either by local or foreign loans or local or foreign investments. For privately owned utilities equity finance is important and for both types of utility, self-financing also. Financial viability is best measured by an annual rate of return on assets, turnover or investment, with some measure of debt service.

References and further reading

1 ROTH, G. (1987) 'The private provision of public services in developing countries' (Oxford University Press) p. 83 et seq.
2 BERRIE, T.W. 'Privatisation in developing countries', *The Crown Agents Review*, **1**, 1989, pp. 27–32
3 LITTLECHILD, S. (1987) 'Privatisation principles', paper to *Seminar on Energy Privatisation* (University of Surrey, Guildford, UK); also, Beesley, M. and Littlechild, S. 'Privatisation: principles, problems and priorities', *Lloyds Bank Review*, July 1983, pp. 7–10
4 BERGE, E. and SHIRLEY, M. (1987) 'Divestiture in developing countries', World Bank Discussion Paper No. 11, Washington DC USA; also, Shirley, M. 'The experience with privatisation', *Finance and Development*, September 1988, pp. 34–35
5 US GOVERNMENT (1978) 'Public utilities regulatory policies act (PURPA)', passed to increase efficiency and competition in the electricity sectors, and to increase the use of cogeneration, autogeneration and renewables
6 PRYKE, R. (1981) 'The future of the electrical industry': internal paper of University of Liverpool; also, Berrie, T.W. 'The new electricity planning', *Energy Policy*, October 1988, pp. 453–456; also, annual operations evaluations reports of the World Bank from 1980; Dutton, P. *et al.* (1989) 'Operation of the new grid company', EE3 Group Project (Department of Electrical Engineering, Imperial College, London)
7 NANKANI, H.B. 'Lessons of privatisation in developing countries', *Finance and Development*, March 1990, pp. 43–45; also, Nankani, H.B. (1989) 'Techniques of privatisation of state owned enterprises, Vol. II (World Bank, Washington DC); also, Vulysteke, C. (1989) 'Selected country experience and reference material, Vol. I' (World Bank); also, Caneloye-Sekse, R. (1989) 'Inventory of country experience and reference materials, Vol. III' (World Bank)
8 MORGAN, R. (1989) 'Prospects of independent power generation', talk at Conference on same title (Surrey Energy Economics Centre, Guildford, UK)
9 Overseas Development Institute. (1986) 'Privatisation: the developing country experience: Briefing paper'
10 CORDUKES, P.A. 'A review of legislation of the power sectors in developing countries', Energy Series Paper No. 22 (Industry and Energy Department, The World Bank); also, Brown I. (ed) (1986) 'Lessons from America, No. 5: the Regulation of Gas and Electric Utilities in the USA' (The Association for the Conservation of Energy, London); also, Major Electricity User Group, (1990) 'Light handed regulation in a restructured electricity industry' (Wellington, New Zealand)
11 WARREN, A. 'Regulating for efficiency', *IEE Review*, November 1990, pp. 372–373
12 *World Bank News*, 4 May 1989 p. 4; also World Bank (1989) 'Developing the private sector: a challenge for the World Bank Group'

Chapter 6
Pricing

6.1 Prescribed pricing

6.1.1 *Introduction*

Traditionally, electricity prices have been set well in advance of usage [1]. Investment and prices are directly related, even though investment is long-term and pricing is short-term. One traditional electricity pricing goal has been to encourage national economic growth using efficient allocations of national resources and ensuring electricity usage is optimal, within consumers' willingness to pay [2]. A second traditional pricing goal has been to ensure consumers are treated equably, another to ensure utilities are financially viable, and another to achieve stability, simplicity and avoid frequent changes in pricing. To achieve these objectives, pricing must be understandable and consumers properly metered, billed and paid up. Used properly pricing fosters national, sector, urban, rural, and similar socio-economic objectives. Considering all the above, practical guidelines for pricing objectives are:

(*a*) Price firmly in accordance with costs
(*b*) Base prices on economic efficiency, costing resources in economic not financial terms [3]
(*c*) Encourage commercial viability in utilities
(*d*) Ensure equity between consumers but with life-line tariffs in development areas
(*e*) Ensure simplicity and comprehension in tariff format.

6.1.2 *Accounting cost pricing*

Averaged historic accounting utility costs [4] are the traditional tariff base, sometimes not even allocating costs between consumer types or classes. However, when incorporating incentives and deterrents, more complex tariffs of this type provide powerful tools for encouraging growth within certain goals (e.g. utility viability, national and electricity growth, discouraging waste, fostering efficiency) although the utility, consumers and national interest often conflict. Complex problems of analysing costs for pricing require many disciplines, and no costing allocation is ever altogether rational or completely fair. In the 1990s, utilities globally will be operated increasingly for profit and will, therefore, gradually waive many traditional objectives, fixing prices more on what the market dictates. Traditionally, utilities have provided and consumers have been charged for both:

(*a*) Actual kWh consumed, i.e. kWh energy charges

(b) Utility always having adequate capacity to supply, even at peak, not much electricity being economically storable, i.e. kW capacity charges. (Predominantly hydro systems make special charges for having water always available.)

The kW capacity service is continuous, not just at peak, and can be costly. To match the method of charging, utility costs are classified in two components: variable, representing kWh supplied; and fixed, representing the kW ability always to supply. For some consumers kW costs exceed kWh costs. The two different types of provision suggest two-part tariffs, with variable kWh components, mainly fuel plus some plant maintenance, and fixed kW components, mainly capital costs allocated to consumer classes and types, voltage levels, etc.

Resolving costs into variable and fixed components is still at the heart of the most advanced pricing and must properly be understood before dynamic pricing, which seems to have no kW charge, is understood (see later). The resolution in Figure 6.1 is for single consumers, consumer groups or utilities. Total supply costs, represented by the line AB, not passing through the origin, are not directly proportional to kWh supplied, the irreducible minimum F representing the fixed capacity costs, the variable cost component V being directly proportional to kWh supplied U;

$$C = F + V \quad \text{or} \quad C = F + vU$$

where v is the variable cost per kWh, $\tan \alpha$.

The terms C, F, V and U can come directly from the utility accounts, past, present, or forecast, quarterly or monthly, but they must all be consistent; V will not always be independent, e.g. with kWh supplied at different times by different generators. A flat tariff per kWh supplied, i.e. with no kW component, historically common worldwide, is still often used in developing countries. Unfortunately, such a tariff can be interpreted as implying the same supply costs regardless of when and where kWh are supplied, which is far from true (see Figure 6.2). Using the symbols of Figures 6.1 and 6.2, average kWh costs are:

$$\frac{C}{U} = \frac{F}{U} + \frac{V}{U} \quad \text{or} \quad c = f + v$$

where c, f and v are total, fixed and variable costs per kWh respectively. The value of C given by $\tan \theta$ in Figure 6.1 varies with U (see Figure 6.2), a hyperbola asymptotic to the vertical axis and to the abcissa, equivalent to a cost per kWh of v, and the fixed to variable kWh costs proportion is much greater for small than large kWh. A flat tariff contains no indication of what the tariff would be with different kWh consumption. The line AB in Figure 6.1 extends to the right up to a value of U for a 100% load factor. Further kWh are then suppliable only by increasing kW, raising the line AB. Similarly, the Figure 6.2 curve extends to the right up to a finite point.

Some fixed cost components do not vary with kW, e.g. for providing consumers' accounts, supplying, installing and reading meters. The costs incurred regardless of kWh consumption or of kW demand can be expressed in terms of total consumers in number, i.e. a consumer component, giving a three-part tariff. Large consumers will have negligible consumer components but

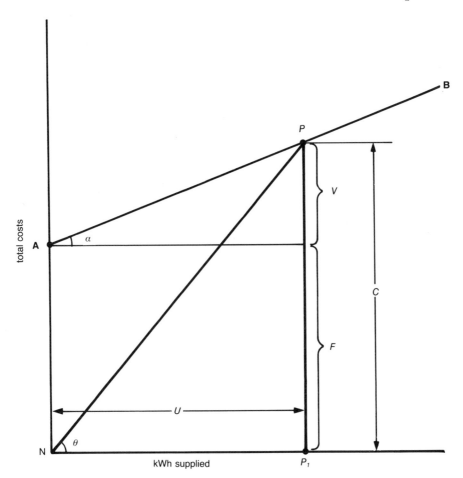

Figure 6.1 Two-part costing (Source: 'Electricity cost and tariffs: a general study', UN, New York, 1972. Reproduced by permission from the UN)
$\tan \theta = $ total cost/kWh $= c$
$\tan \alpha = $ variable cost/kWh $= v$
$F = $ fixed costs
$V = $ variable costs

small consumers may have high components, which may justify charging small consumers a higher tariff per kWh than larger consumers.

Three-part tariffs are illustrated by Figure 6.3. The horizontal axis OX shows kWh supplied, the other OZ shows the kW peak demand; the vertical axis OY shows the total costs. For combinations of kWh and kW, the total costs are a plane surface FDEB inclined at an angle α to the horizontal in one direction and at an angle β in the other. For a kWh consumption OU and kW ON, total costs are shown at P, lying vertically above a point P_1 having ON and OU for horizontal coordinates. The height PP_1 represents total:

154 *Pricing*

C = consumer component ⎫ The two combined
M = kW component   ⎬ form fixed component F
and V = variable component ⎭

Per kWh, the variable component is tan α, the kW component tan β and the average total cost tan θ. The plane APP_1N is a repetition of Figure 6.1. Figure 6.3, like Figure 6.1, can represent single consumers, consumer groups or whole utilities.

6.1.3 *Marginal cost pricing*

As an alternative to using utility accounting costs, incremental or marginal supply costs can be used and the previous analysis repeated, but this time using incremental costs for supplying an additional kWh theoretically at any instant

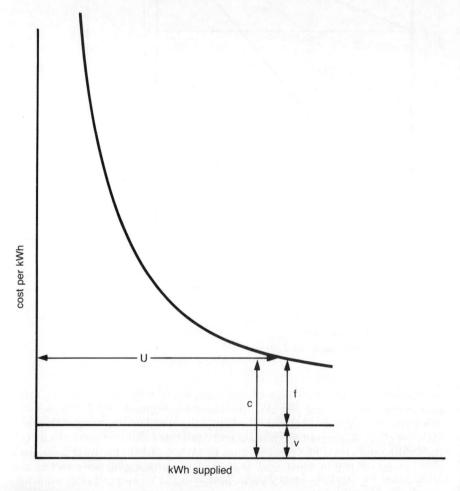

Figure 6.2 *Average cost per kWh (Source: 'Electricity cost and tariffs: a general study', UN, New York, 1972. Reproduced by permission from the UN)*

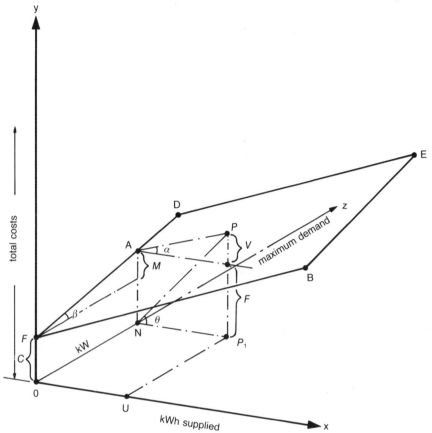

Figure 6.3 *Three-part costing (Source: 'Electricity cost and tariffs: a general study', UN, New York, 1972. Reproduced by permission from the UN)*
C = consumer costs
M = MD costs
F = fixed costs
V = variable costs
$\tan \alpha$ = variable cost/kWh
$\tan \beta$ = fixed cost/kW
$\tan \theta$ = average cost/kWh

in time, i.e. the difference in supply costs for supplying, compared with non-supplying, the additional kWh at any instant in time. The first stage in such an exercise is to determine prices strictly in accordance with economic efficiency, i.e. strictly from the marginal costs as these are measured by the national economy. In the second stage each of the objectives and guidelines mentioned earlier must be carefully looked at in turn and the efficiency prices adjusted accordingly. The third stage is to modify these adjusted prices to ensure utility financial viability, e.g. with respect to cash flows and revenue requirements [5].

The same multi-part tariffs as for accounting cost pricing can be applied to marginal costs, these tariffs being determined theoretically for each point in time and in practice averaged over various time-spans. This usually necessitates a computer, especially when tariffs are being determined from long-run marginal costs of a utility's least-cost development programme [6]. Until the late 1980s this was regarded by most utility planners as the optimum.

To cover all pricing objectives mentioned earlier, marginal cost prices should at least start off on a resource or opportunity–cost basis, i.e. in terms of alternative economic benefits foregone by consuming incremental resources now rather than later, especially applicable to scarce capital or fuel resources. Even if an economy possesses prodigious fuels, the acceptable policy must still be to keep to strict marginal opportunity-costs of fuel supply when pricing. Otherwise, energy sector prices will never become related to import or border fuel prices, or to costs of developing alternative fuels, or to general opportunity-costs in an economy. An economy might possess only small amounts of fuel for producing electricity; in this case, what resource cost to use for pricing is almost irrelevant, the economy only able to benefit for a short time from the small amount of indigenous fuel, after which only imported fuels can be planned for, more expensive than indigenous. In the more normal cases between these two, electricity pricing should be adopted to take account of using indigenous fuel resource over a specified time, while still remaining as economically efficient as possible. One way in common use:

(a) Looks at the depletion time of fuel source in question, say 20 years, then compares this with the expected time, plus annual costs, of building up alternative energy sources; then
(b) Calculates the difference between the annual cost of alternative energy sources and the annual cost of extraction of the indigenous fuel; then
(c) Discounts all annual costs back to a present value using the discount rate appropriate for fuels in a particular economy; then
(d) Sums the discounted values; then
(e) Adds to the extraction cost the depletion premiums, found by the above described discounting and summing up process; and takes stock.

There is thus a determinable price to be paid today for using indigenous fuel now to produce electricity rather than later, often called a 'user price' or a 'rent', or a 'depletion premium', which must be allowed for when adjusting theoretical economic-efficiency derived marginal electricity prices for national or utility fuel policy.

6.1.4 *Peak and off-peak*

The concept of multi-part tariffs, introduced earlier, is now extended for incremental kWh varying from one moment to the next, identifying the relevant marginal costs. On a strict incremental basis, theoretically all capacity costs should be borne by peak consumers, who should also bear the variable costs at peak; off-peak consumers would then need to pay only for variable costs, leading directly to time-of-use pricing, described later. Yet, once capacity is operational, it is available for use peak and off-peak; thus, what is needed is to first determine the optimal capacity to provide for peak and off-peak taken together, then to determine the cost allocation for time-of-use.

The national, and to some extent the utility's objective, is maximising net social economic benefit, i.e. total social benefits less total social costs. A different analysis would follow for optimising completely with respect to the utility or the consumer, but the arguments and formulae would be similar, but using accounting rather than economic costs. Assumed are distinguishable demand in two equal periods, e.g. night and day, winter and summer, each period demand depending only on prices charged in that particular period, i.e. assuming independent demand against price curves for each and every period. In the first considered case, a firm peak demand of only one period causes capacity additions, and charging different prices, peak and off-peak, leaves the peak uneliminated. In the second case the peak moves, and the demands of both periods are responsible for new capacity additions, and the charging of differential tariffs can result in the previous off-peak period becoming the peak period. In Figure 6.4 showing the economists' price against supply cost 'demand' curves [7], day and night electricity demands, each of twelve hours duration with no variation within those twelve hours, are shown as $D_d D'_d$ and $D_n D'_n$, respectively. Inherited capacity has a constant per unit assumed variable cost of Ob, with capacity of a rigid output limit of Od in each period. The pricing policy leading to optimum use of inherited capacity gives appropriate prices of P_n and P_d for night and day output respectively. Charging these prices would result in an aggregate social benefit of $bD_n P_n + bD_d P_d A$, the maximum

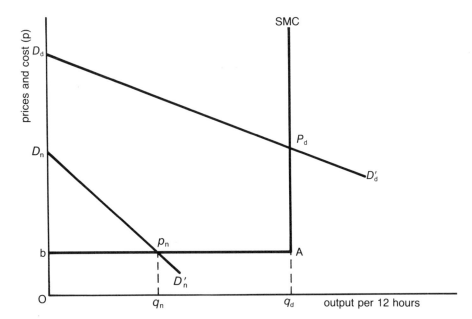

Figure 6.4 *Price and cost against output (Source: HIRSCHLEIFER, J. 'Peak loads and efficiency pricing: comment', Quarterly Journal of Economics, August, 1958, Figure 8.1. Reproduced by permission of the Journal)*

158 *Pricing*

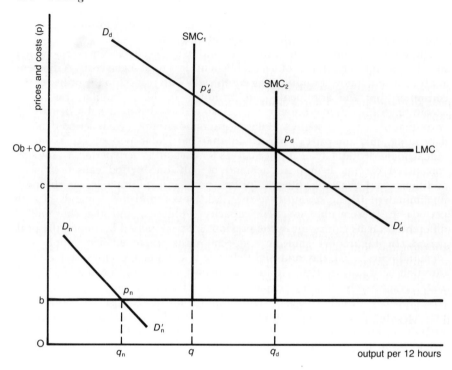

Figure 6.5 *Firm peak case (Source: HIRSCHLEIFER, J. 'Peak loads and efficiency pricing: comment', Quarterly Journal of Economics, August, 1958, Figure 8.2. Reproduced by permission of the Journal)*

obtainable. Outputs in each period are q_n and q_d for night and day respectively. The day price will act as a rationing device allocating available output among potential consumers. The concept of marginal costs appropriate to the day price is that of opportunity-cost already described.

Determining the optimum capacity involves a consideration of short-run (SMC) and long-run marginal costs (LMC) and when they are equal. To retain two-dimensional presentation in Figure 6.5, assumptions are made: no inherited capacity; no capacity indivisibilities; no technological changes; plant life given and constant; demand conditions constant; plant variable (running) costs the same for each year throughout life; and capital costs as annuities. With constant running and capital costs at Ob and Oc per unit, respectively, then a capacity limitation short-run (SMC) is Ob and long-run (LMC) is Ob+Oc. For simplicity, only two possible output scales are shown, producing outputs q and q_d, respectively, with short-run marginal costs SMC_1 and SMC_2. With capacity size q constructed, the two period prices would be p_n and p_d, while with capacity size q_d constructed, the two period prices would be p_n and p_d. Capacity q_d is in fact optimal; off-peak consumers pay price Ob, peak consumers price exceeding Ob, by sufficient to cover the marginal cost of extra capacity unit Oc. When capacity with an output exceeding q_d is in operation, the price paid during peak period would fall short of that required to justify the last unit of capacity constructed, i.e. while this price would exceed Ob it would be less than Ob + Oc. With capacity of output less that q_d in operation, marginal addition to capacity is justifiable since the peak period price paid would have exceeded Ob + Oc. In this firm peak case all capacity charges (Oc) (q_d) are borne by peak consumers. Charging differential prices does not result in off-peak exceeding peak demand and therefore requiring more capacity than is available. Output capacity is determined for peak period consumers' demands, being a 'free good' to off-peak consumers. The previous conclusions for Figure 6.4 thus hold for a firm peak and 'rigid' plants.

Figure 6.6 again assumes rigid plants with night demand price of Ob, capacity required is q_n; with day price of Ob + Oc, available capacity would be q_d. However, this means the off-peak price is charged to that demand which made the peak demand on capacity, and the off-peak demand becomes the peak demand. In this case of shifting the peak, the economic efficiency solution is found from demand curve Dc, an aggregate demand curve obtained from summing day and night demand curves, which can be added directly because any unit of capacity can be justified by either demand in one period alone, or by combined demands of two or more periods. Since these demands are complementary not competitive, vertical summation is required. Using this combined demand curve, optimal capacity is determined by the intersection of Dc with the LMC (joint) curve, showing the cost of meeting a unit increase in demand

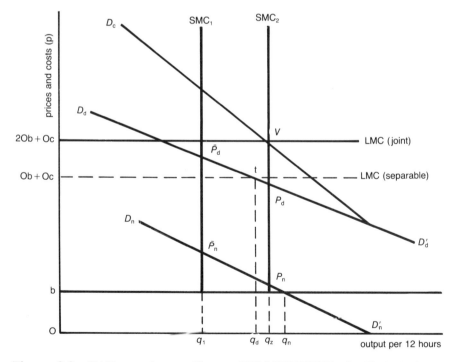

Figure 6.6 *Shifting peak case (Source: HIRSCHLEIFER, J. 'Peak loads and efficiency pricing: comment', Quarterly Journal of Economics, August, 1958, Figure 8.3. Reproduced by permission of the Journal)*

over the 24-hour demand cycle, made up of the sum of the unit running costs of capacity in each of the periods, which is 2Ob, plus the cost of providing an additional unit of capacity for the whole demand cycle, which is Oc. As with a firm peak, costs of providing a unit increase in output within either period alone when capacity extensions are required is Ob + Oc, shown in Figure 6.6 as LMC (separable), with optimal capacity q_z. Output in each period should be extended to use this capacity providing the running costs of such output are covered. The optimum prices to charge are then $q_z p_n$ at night, and $q_z p_d$ for day output. The sum of these prices equals $q_z V$, the long-run marginal cost of providing an additional unit of output in both periods, i.e. the price charged in each period is different but the output is the same.

6.1.5 *Time-of-use pricing*

Time-of-use pricing is but an extension of the above type of analysis. Many wholesale or bulk supply tariffs (BST) employ marginal cost pricing [8], identifying expenditure on costs related to consumers (kW), capacity (kW) and energy (kWh) as above, this time referring to distribution utilities, consumer groups or very large consumers. BST also normally use time-of-use pricing. Merit order (Chapter 7) of generator operation ensures that short-run marginal costs of kWh supply increase with loading. Figure 6.7 shows a typical winter loading curve for which marginal costs in the afternoon are higher than at night. Having enough capacity to meet the peak means some generators being utilised less than others, suggesting different generator types be installed depending on the required utilisation, and the marginal capacity costs to meet a kW change can shift the whole load curve up or down, or at least change its shape so that it becomes smoother or peakier. One pricing strategy to recognise this distinguishes kW charges between capacity required to meet the extreme annual peak(s) and that needed for the winter (or summer) 'plateau', the basic loading level of each day, between 8.00 am and 4.00 pm. On this basis is calculated the relevant marginal costs. Basic demand permanently shifting upwards requires new plant; a permanent shift downwards means deferring new plant. Any new plant for plateau use will be the most modern, highest efficiency type; in the past base-load steam plant, although new types of plateau plant are rapidly coming (Chapter 7). The appropriate marginal kW cost to use is the net effective cost (NEC), an estimate of the effect of investing in one new plant on the total system costs taken over the planning horizon, i.e. the summed net present value of all the annual capital costs of new plant and the annual system running costs, over the planning horizon, including cost for incremental transmission. Table 6.1 shows a calculation of this marginal cost; the basic capacity cost was then being set at UK£30 per kW.

The difference between actual system peak and basic winter/summer plateau forms the 'peaking' demand. As the system approaches an optimal long-run plant mix, any incremental peaking demand will be met most economically by installing gas turbines, and the NEC of this plant is the marginal cost of incremental peak demand. However, with energy conservation measures, peak demand will fall permanently; it will then be economic to decommission older plant which is only used to meet peak demand, saving (say) UK£5 to UK£10 per kW on a basis comparable to the UK£30 per kW for a new gas turbine plant

Pricing 161

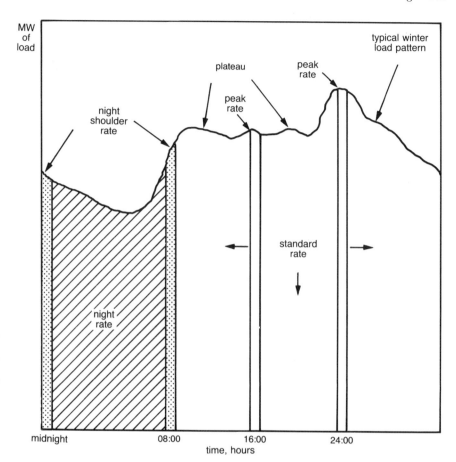

Figure 6.7 *Typical load curve and 1980–81 BST unit rates (Source: 'Central Electricity Generating Board', Report by the UK Monopoly and Mergers Commission, HC 315, May 1981, Figure 1, p 307. Reproduced by permission of the Controller of HMSO)*

(Table 6.1). One method of proceeding is to compromise on an average of (say) UK£14 per kW to represent the marginal peaking capacity cost. Wholesale time-of-use kWh prices are based on the marginal running cost averaged over periods in time at the load levels in question in plant merit order. The day is divided into periods (Figure 6.7) and the marginal costs calculated from the plant last brought into service to meet the average load level in each period, the main component being fuel cost (Table 6.2).

In a like manner, summer and winter kWh prices can be set separately, also week and week-end prices. If the tariffs have to differ from the marginal costs for utility revenue recovery reasons in the short-term, the brunt of the change is best borne by the less elastic features of the wholesale tariffs, particularly the basic capacity charge and one of the standard kWh rates.

6.1.6 Retail tariffs and consumer response

Retail tariffs will normally be based on wholesale tariffs, resembling these in complexity in so far as the country, sector, utility and type of consumers warrant it. To make the best use of marginal-cost pricing adequate metering is essential. Time-of-use metering is best for marginal-cost pricing, providing it is economic. The two-part tariff explored earlier separates kWh from kW charges. With some small metering alterations it can instead separate out two or more different kWh charges. Neither the current limiters nor the interruptible tariffs of Chapter 3 are fully defensible by marginal-cost pricing theory because they both assume the relationships between consumers' peak demand and total kW of connectable apparatus are accurately known, and that, when demand presses on supply, the interruptible tariff increases rapidly to infinity at cut off. This is, in fact, true only for dynamic pricing (see later). In practice, large and medium industrial consumers supplied directly from the grid have for some years paid separate kW and kWh prices, the latter varying by time-of-day or time-of-year. Other consumer classes have been introduced to many other types of tariff structures based on BSTs, e.g. tariffs in the UK and France, where there exists

Table 6.1 *The CEGB's calculation of the marginal cost of new capacity for the 1980–81 BST (Source: 'Central Electricity Generating Board', Report by the UK Monopoly and Mergers Commission, HC 315, May 1981, Table 1, p 309. Reproduced by permission of the Controller of HMSO)*

	Coal (considered to be capacity to meet basic demand) £/kW	GT (considered to be capacity to meet the peaking demand) £/kW	Retaining old plant £/kW
Basic estimate of net effective cost of new plant (incl associated transmission)	24	25	
Allowance for planning margin	+6	+7	
Allowance for transmission losses	+1	+1	
	31	33	5–10
Less correction to bring total revenue down to Government financial target	1	3	
Estimate used for 1980/81 BST	30	30	10

Table 6.2 *Calculation of BST energy charges (Source: 'Central Electricity Generating Board', Report by the UK Monopoly and Mergers Commission, HC 315, May 1981, Table 2, p 310. Reproduced by permission of the Controller of HMSO)*

Period	Estimated Marginal Costs Price* (pence per kWh)	BST rates including 1980 est of fuel cost adjustment	Estimated BST rates as a proportion of marginal costs
Peak	4.97	4.33	87%
Standard	2.70	2.41	89%
Night	1.32	1.34	102%
Shoulder	2.23	2.02	91%

* Assuming national fossil price of 4.250 pence per tonne (i.e. the forecast for 1980–81 made in March 1980 when the tariff was set).
These figures are based on estimates of output related works costs (other than fuel) at the marginal plant of:

Period	Costs p/kwh
Peak	0.5
Standard	0.3
Night	−0.05 (i.e. the CEGB saves)

substantial thermal generation, reflecting how crucial it is to use time-of-use and not just two-part marginal costs kWh tariffs, especially at times of peak and the shoulders of the peak. Where hydro plant dominates, as in Scandinavia, seasonal variations in water availability (e.g. in the winter months) and the additional cost of storage, dictate both wholesale and retail tariff structures, and each country's philosophy is different, there being no general rules for pricing.

Experience in Europe and the USA indicates considerable benefit to implementing marginal cost pricing in all its forms to all industrial and commercial consumers. Typical changes in daily load curves were [9] illustrated by 515 large industrial and commercial consumers in five major electricity service areas of the States of California, Michigan, New York, and Wisconsin, having minimum peak demands ranging from 300 kW to 5 000 kW, when responding to TOU pricing as follows:

(a) Approximately 20% of total consumers reduced peak demand by shifting to off-peak usage; the remainder do so without shifting to off-peak usage but by using more efficient equipment for lighting, ventilation, air conditioning, etc.

(b) Approximately 55% of total consumers did not change peak demand; about 82% of these did not respond to TOU tariffs because of inflexible production processes, or because changing to off-peak tariffs, all other factors being included, would not be cost-effective; only about 18% of the 55% made no effort to change their production habits.

(c) Approximately 26% of total consumers increased their peak demand because they stepped up their production.

164 *Pricing*

(d) Approximately 90% of total consumers demonstrated that they fully understood the TOU tariff they were using or were offered.
(e) Approximately 67% made studies of how new tariff structures being offered would affect their business.

In Europe and the USA consumption for most small commercial and most residential consumers is still done using simple metering and tariffs, because tariffs for these consumers presently cannot be complicated to be understood. This is likely, however, to change in the 1990s, i.e. it will be economic for marginal cost and TOU tariffs to be used for more and more of these consumer classes as cheaper, sophisticated meters become available. Indications on the effect of TOU metering on residential consumers in Arizona, Connecticut, Ohio, North Carolina, Rhode Island and Wisconsin are:

(a) Noticeable reduction in peak kWh consumption
(b) Increase in diversity factor for apparatus
(c) Reduction in shoulders of demand
(d) Little evidence of new peak creation
(e) A ratio of eight to one in the peak to off-peak price needed to reduce average peak to off-peak consumption during the summer by 24%.

The empirical evidence from outside Europe and the USA is that, where marginal cost tariff structures have existed for sufficient time to become effective, say for five to ten years, they have done a lot of the job for which they were instituted, at least for industrial and commercial consumers. For example, most industrial and large commercial consumers in developing countries have responded well to peak load pricing by: changing production patterns over time; using off-peak periods more and peak periods less; generating themselves if they are co-generators or autoproducers, when practical and economic to do so, especially during times of peak demand; and physically storing energy, often as heat, extracted off-peak for use at peak. What is happening in developing countries under marginal cost pricing is large load shiftings from peak to off-peak periods but not peak demand reduction, compared with the peak loads of European industries reducing over a range of 30% to 90% between 1960 and 1975. In developing countries and in some developed countries, reduction in peak demand has been accompanied by increases in annual load factors. Residential consumers appear to respond less well than industrial or commercial consumers, to marginal cost and TOU pricing because of lack of consumer understanding. On the other hand, there has been enthusiasm for the introduction worldwide to residential tariffs of simple off-peak tariffs, e.g. for water heat storage, and heat or cool space storage, and for specific residential loads centrally turned off and on by remote control, under DM schemes.

6.1.7 *Commentary summary*

Traditionally, electricity prices have been prescribed well in advance of usage. Following on from historic accounting utility costs pricing, incremental or marginal cost pricing grew in importance from the 1960s onwards, referring to incremental kW and kWh demands at any particular instant. Although in the 1990s this approach is being challenged, prices will always be derived at least partly from the planning process of finding the least-cost utility investment programme deriving the marginal cost of supply from the least-cost programme, even if these cannot be used directly but have to be modified for metering practicalities and revenue recovery considerations. Two-part (i.e. kW and kWh) tariffs are much more effective than single value tariffs, leading to both kW and kWh time-of-use pricing, at least for wholesale electricity sales.

6.2 Improving markets

6.2.1 *Sector changes*

In the 1990s electricity sectors worldwide are undergoing rapid, irreversible changes [10]; with volatile fuel costs, unpredictable load growth, privatisation deregulation, conservation, environmental management, and revolutions in technology. Needs for efficiency improvements and increased flexibility to handle uncertainties are stronger and more challenging than ever. Electricity worldwide is a billion dollar industry causing governments concern about utility-consumer co-operation, innovative tariffs, and better transparency on utility and consumer costs.

6.2.2 *Criteria*

Any change in pricing to improve electricity marketplaces must be built round four basic criteria:

(1) Economic efficiency; encouraging consumers to adjust usage patterns to match utility marginal costs, subject to revenue reconciliation and transactions costs
(2) Equity; reducing consumer cross-subsidies, making consumer charges near to utility supply costs
(3) Free choice; providing consumers with options on cost and supply standards, how they may choose to use electricity
(4) Utility control; operation and planning, considering the engineering requirements for controlling, operating and planning.

Most 1980s electricity pricing fell short of meeting these criteria, sometimes to a large extent. Prescribed marginal or time-of-use tariffs proved not directly related to actual marginal costs at all. Demand (kW) or fixed charges, especially consumer related charges, remained a constant anachronism and source of challenge by consumers, these having only a tenuous relationship with marginal costs, often causing counter-productive consumer–utility relationships. Demand management often involved direct utility control which could foster wasteful consumer behaviour. Cross-subsidisation between consumer classes, and even within a given class, was rampant. It needs to be realised, however, that the criteria above cannot be achieved by just adding the outputs from modern micro-electronic devices to traditional pricing. It is only by viewing the systems of utilities and consumers as one integrated system, with market clearing prices devised accordingly and determined only by supply and demand (similar to commodity markets), that these criteria can be properly met.

6.2.3 Market transactions

All successful markets need:

(a) Supply sides with costs increasing with demand
(b) Demand sides with demands adapting to price
(c) Market mechanisms for buying–selling and for clearing the market efficiently
(d) No monopolistic demand behaviour
(e) No monopolistic supply behaviour

Electricity is ideal for the first four of the above needs:

(a) Figures 6.8 and 6.9 illustrate changing supply costs with supply-demand conditions, Figure 6.8 marginal fuel costs, Figure 6.9 total marginal costs, including quality of supply (reliability) premiums (see later), and any revenue reconciliation costs needed, realising that such marginal cost variations are highly correlated with demand levels.

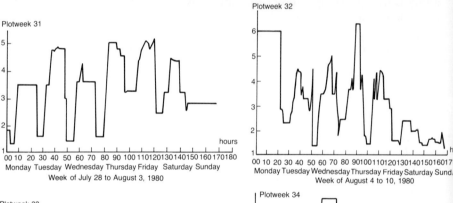
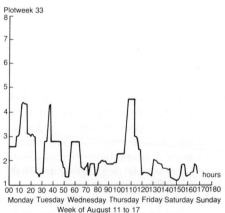
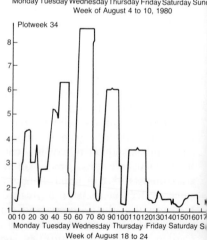

Figure 6.8 *Measured marginal fuel costs (Source: SCHWEPPE, F.C. et al, 'Spot pricing of electricity', 1988, Kluwer Press, Norwell, Massachusetts, USA, Figure 1.1.1. Reproduced by permission of Kluwer Academic)*

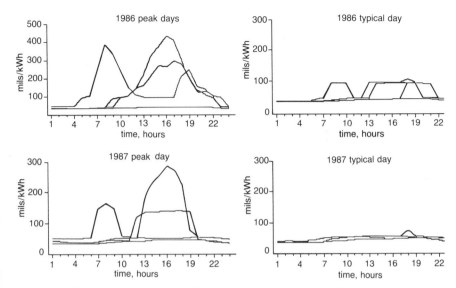

Figure 6.9 *Predicted marginal fuel, capital and other costs (Source: SCHWEPPE, F.C. et al, 'Spot pricing of electricity', 1988, Kluwer Press, Norwell, Massachusetts, USA, Figure 1.1.2. Reproduced by permission of Kluwer Academic)*

(b) Existing market mechanisms, e.g. billing, are easily adaptable.
(c) Industrial consumers with demand meters need no new hardware to implement prices varying each hour.
(d) Monopsonistic behaviour is difficult demand-wise because of the large number of consumers.

For the fifth need above, marketplaces most likely to exist in the 1990s may well have a mixture of independent consumers and heavily regulated privately owned or government owned utilities.

Up to the 1980s, markets were designed explicitly around electricity control and operation, meeting the four basic criteria above. But they would not automatically do so under a change to pricing where all utility–consumer transactions are based self-consistently on one single quantity, i.e. the hourly 'spot' price as determined by the actual demand (i) at that hour; (ii) the actual hourly variable cost; and (iii) the actual capability of supply of generation, transmission and distribution. To get such a set of arrangements to work, there must be feedbacks between utilities and consumers, and between generation, transmission, distribution, and consumers, but without removing freedom of choice for all concerned. Benefits from balanced real-time utility-consumer feedback marketplaces are as large as transaction or communication costs conveying the needed information. With respect to the market mechanisms, some consumers (e.g. large industries) would then see prices updated (say) each hour, while others (e.g. residences) would see prices updated only each billing

168 *Pricing*

period, unless consumers want to exploit prices set hourly, then paying the additional transaction costs to do so, but repaying the 'spot' price benefits. Such 'spot' pricing is now explored further.

6.3 Dynamic, spot and real-time

6.3.1 *Differences*

A common misunderstanding is that the major difference between dynamic and the still omnipresent prescribed pricing, is in dynamic prices having complicated time-of-use variations. As described earlier, prescribed pricing from the 1970s has increasingly had complicated time-of-use variations. The crucial differences between the two types of pricing lie in the following: (a) the nature of the price variations and not their presence; and (b) when the pricing is applied with respect to electricity consumption.

Figure 6.10 plots historical variations of prices in US¢ per kWh for one utility. The hourly cost, equal to the spot price variations of Figures 6.8 and 6.9, simply

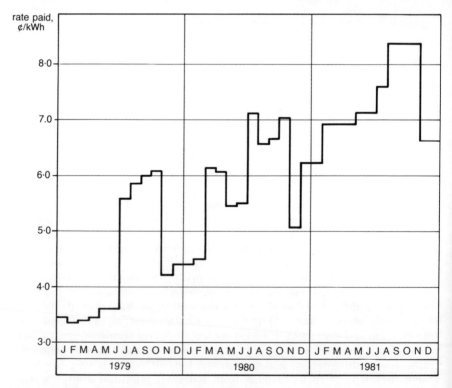

Figure 6.10 *Examples of monthly variations in residential prices (Source: SCHWEPPE, F.C. et al, 'Spot pricing of electricity', 1988, Kluwer Press, Norwell, Massachusetts, USA, Figure 1.1.3. Reproduced by permission of Kluwer Academic)*

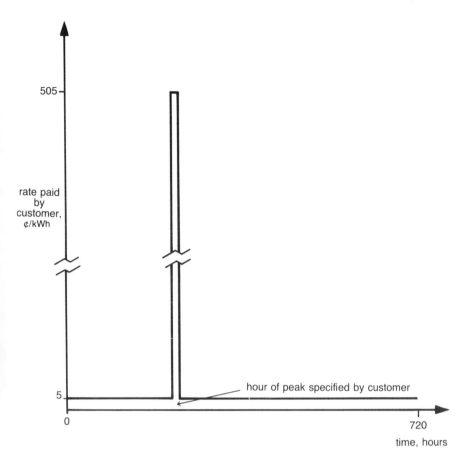

Figure 6.11 *Effective price variations from demand charge (Source: SCHWEPPE, F.C. et al, 'Spot pricing of electricity', 1988, Kluwer Press, Norwell, Massachusetts, USA, Figure 1.1.4. Reproduced by permission of Kluwer Academic)*
energy charge = 5¢/kWh
demand charge = 5$/kWh

exhibits finer time detail. An industrial consumer has a (say) US¢5/kWh energy charge and a (say) US$5/kW demand charge based on electricity used during the consumer's peak hour during the month. The corresponding kWh rate paid by the industrial consumer is plotted in Figure 6.11 displaying a dramatic time variation bearing little resemblance to either Figure 6.8 or 6.9. A second difference is in the relationship between utility and consumers, i.e. unlike prescribed markets, those for spot prices have utilities and consumers as partners, working together for their maximum joint benefit from electricity usage at minimum cost, and this type of partnership needs building up from scratch if spot pricing is to work well.

170 Pricing

As stated earlier, despite major differences of concept, spot and prescribed pricing are based on the same basic principles and an understanding of prescribed pricing is needed to understand both. Prescribed flat and time-of-use tariffs are specified from expected costs, rates of return on assets, etc.; and hourly spot price is based on the same principles, but with spot pricing, the price is worked out every hour by computers instead of once a year in advance. Existing direct demand control and interruptible tariffs are special cases of marketplace transactions, and electricity system operation has often included a real-time spot market for purchases and sales between utilities since the 1970s. The spot price markets of the 1990s simply extends this former type of spot market to include consumers as well as utilities, being the logical evolution of prescribed pricing, DM, improving efficiency, conservation, etc., carried out previously within the established practices of system operation, given the concept of utility-consumer partnership and availability of inexpensive communications, control and computation equipment. A spot price market meeting the original four criteria is best achieved in three steps:

(1) Define hourly spot prices and evaluate their behaviour
(2) Specify an appropriate set of utility–consumer transactions based on the hourly spot price and associated transactions' costs
(3) Implement the electricity marketplace, considering the needs and capabilities of both utility and consumers.

Owing to their importance for the 1990s these are now briefly considered in turn.

6.3.2. Step 1: *Define hourly spot prices*

An hour as the fundamental time unit is convenient but not essential; it could range from minutes to several hours. The hourly spot price is defined in terms of marginal costs, subject to a quality of supply premium to ensure demand does not exceed supply, and revenue requirements. It depends on hourly variations of:

(a) Generation fuel costs and capacities
(b) Transmission and distribution network losses and capacities
(c) Aggregated consumer demand patterns.

The hourly spot price itself is random (see Figures 6.8 and 6.9). Its future value cannot be forecast exactly, due to random equipment outages and random demand variations. An hourly spot price can be determined for a utility which is buying from, as well as selling to, its consumers. The buyback hourly spot price can be either greater or less than the selling hourly spot price.

6.3.3 Step 2: *Specify utility–consumer transactions*

All utility–consumer transactions will be based on the hourly spot price as defined in Step 1. The three general types of transactions are:

(a) Price only
(b) Price-quantity
(c) Long-term contracts

Examples of price-only transactions are:

Pricing 171

(a) One-hour update: consumers see prices (US¢kWh) varying for each hour, predicted and communicated one hour in advance
(b) 24-hour update: consumers see prices (US¢kWh) varying for every hour, predicted and communicated 24 hours in advance
(c) Time-of-use (TOU) and flat rates: tariffs (US¢kWh) are calculated using predictions of hourly spot price behaviour, one billing period in advance; these marketplace TOU and flat rates differ from prescribed ones introduced in the 1970s because they vary each billing period and never contain a demand charge, the latter being absorbed in the quality of supply premium which is charged at any hour to prevent demand from exceeding supply.

The choice of what type of price-only transactions to offer is based on trade-offs between the cost of transactions, metering, communication, computation, etc., and the benefits obtained from transactions, e.g. large industrial consumers might want one-hour update prices while small residential consumers might want flat rates.

Price–quantity transactions involve a short-term utility–consumer contract, e.g. a consumer may choose to receive a specified quantity of electricity at a lower price, with a contract to drop all or a part of such usage on a utility signal. Price–quantity transactions include as special cases, prescribed interruptible contracts, and some direct demand management. However, spot price markets transactions are in themselves more robust and by their very nature can better match consumers' needs and capabilities.

Long-term contracts are fixed price, fixed quantity. They can extend days, months, and years into the future, e.g. assume that on 1 January an industrial consumer contracts for 1000 kWh of energy to be delivered between 10 and 11 am on 1 July for US¢10/kWh. If, when 1 July finally comes, the price between 10 and 11 am is actually US¢9/kWh, the actual cash flow is as follows:

If consumer uses	The consumer pays
1000 kWh	US$100
2000 kWh	US$100 + US$90 = US$190
0 kWh	US$100 − US$90 = US$10

Such transactions enable a consumer to buy an insurance policy by locking in 1000 kWh at US¢10/kWh, even though the consumer still sees the spot price for actual usage.

6.3.4 *Step 3: Implement marketplace*

Full implementation of the spot market requires the utility and its consumers operating as partners. Utility implementation means making real-time calculations and predictions of hourly spot prices, metering, communications, billing and system operation, using the new prime control signal called price. The impact of the spot price market on utility long-term investment decisions then becomes important. Consumers choosing to exploit market potentials are concerned with implementing appropriate response systems ranging from simple manual responses to sophisticated digital-device control. The utility can

172 Pricing

provide the control mechanism as a consumer service. The next two sections provide examples of spot price market implementation, so important for the 1990s.

6.4 Spot market implementation

6.4.1 *Developed country*

Figure 6.12 shows the transactions and information flow of a spot market in a developed country, suitably simplified to show only one consumer. Each box is now briefly considered. With respect to the generation, transmission and distribution box, the spot market impact on capacity requirements, generation mix, etc., is long-term, often resulting in reduced capacity per demand unit in a

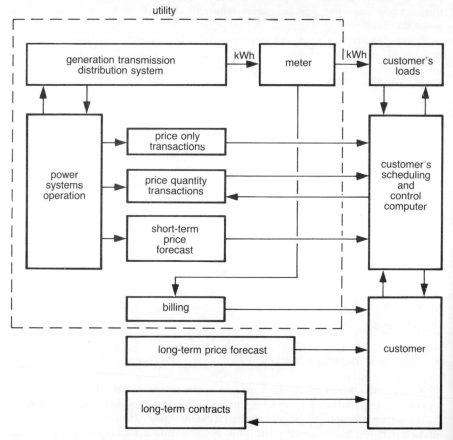

Figure 6.12 *Functions and information flow in energy marketplace (Source: SCHWEPPE, F.C. et al, 'Spot pricing of electricity', 1988, Kluwer Press, Norwell, Massachusetts, USA, Figure 1.3.1. Reproduced by permission of Kluwer Academic)*

changed mix. All traditional basic items in the system operation box continue under spot pricing with modifications, e.g. short-term demand forecasts now include price effects, plant commitment logics incorporate price feedbacks, spinning reserve is carried by demand as well as supply, whichever the cheapest. In the meter box a recorder measures, then stores, hourly electricity usage for each consumer, read once a month and used to compute monthly bills.

For the price-only transaction box, at (say) ten minutes to each hour, the one hour update spot price is computed from a one hour forecast of system conditions, fuel costs, losses, reserve margins, network capabilities and demand, including price effects. Automatically this one hour update is sent digitally by telecommunications circuits to consumers' computers connections, holding for one hour. At (say) 3 pm the hourly spot prices for the 24 hours, starting at 3 am next morning are computed using operating conditions forecasts. Consumers on 24-hour updates can receive these 24 numbers by telephone at any time after 4 pm or by consulting newspapers, radios or TVs. In the price–quantity box, transactions are available only to consumers on a 1-hour update basis. Ten minutes before each hour, an interruptible spot price for the next hour is computed based on system conditions and forecasts of how much interruptible electricity will be purchased by consumers and telecommunicated to those concerned, consumers wanting interruptible energy communicating their secure level, all usage above that level being at interruptible price. In short-term emergencies, e.g. generator or major transmission forced outages, the reduction required from each consumer is computed and telecommunicated to the consumer's computer. In the short-term box, forecasts of hourly spot prices for each hour for the next week are available by consumers' telephone, either verbally or digitally, but these forecasts can be wrong. In the billing box, bills are computed each month by summing metered and recorded hourly electricity use times and spot prices existing at corresponding hours. For the long-term price forecast box, monthly and yearly price forecasts are made by utilities or independent consultants. Long-term contracts (fixed in price and quantity) are written by brokers who contract directly with consumers, but such transactions have no direct effect on actual system operating.

6.4.2 *Large industrial*

An industrial consumer, with (say): (a) a foundry with many metal-working machines of 1kW to 100kW; (b) electrical metal melting, using 10MW to melt, 1MW to maintain molten; (c) lighting; and (d) office space conditioning with a sophisticated computer system used for production scheduling, hourly for the next day and weekly, using linear programs and direct control of scheduled processes. Consumer interaction with the electricity marketplace is through a plant manager and a production manager, the latter for long-term, monthly to yearly issues, making long-term contracts by combining long-term price forecasts with forecasts of future electricity needs and consumer cash flows. The plant manager is concerned with hourly and daily operation, providing inputs into consumers' control and scheduling computer. Plant operation scheduling is by combining short-term price forecasts with specified product mix to be produced, e.g. metal melting always scheduled for low price times. Some large metal-working machines are scheduled for low price times. The office space conditioner makes use of preheating and precooling the building, so that space conditioning load is lowest during high spot prices and forecasts of future outside temperature are inputs to optimisation. With prices very low (e.g. at night) some metal curing processes using gas are switched to electricity. Price–quantity transactions are adjusted and metal melting is always done on interruptible spot prices. During times of large differences between secure and interruptible spot prices, the secure electricity purchase level is reduced, so that some of the other end-uses come into the interruptible tariff, i.e. unless there is a tight production schedule. All scheduling operations are done automatically by computer using input instructions plus data provided by the plant manager.

6.4.3 *Residential*

Sophisticated residential consumers under 24-hour update have small special purpose computers, automatically getting once each day 24 prices for the next. Computers control the space conditioning and water heating to meet consumer requirements at minimum cost. When prices rise above critical consumer-specified levels, computers warn so that manual control can be used, i.e. computers and expert systems helping consumers diagnose needs, then make rational decisions for price variations. Simple residential consumers also see 24 hour spot prices but have no computer, mostly ignoring price variations, operating routinely. However, at some critical time(s) of year with high prices, consumers exercise manual control, learning that such times exist by reading newspapers and/or listening to utility announcements via radio and TV.

6.4.4 *Developing countries*

Developing countries' electricity is expanding fast, meeting the needs of rapidly changing, unsophisticated societies and these make for important differences from the cases above. In such countries only large industrial and commercial consumers see 24-hour and 1-hour updated spot prices, mostly the 24-hour update. Metered residential consumers see flat rates possibly updated each billing period. Residential metering systems are installed, bearing in mind that developing countries will eventually become developed, and spot marketplaces then fully established, e.g. electronic systems could be employed in the 1990s to avoid later the trap all developed countries face in the 1990s, namely, having large investment sunk in an already supplanted technology (e.g. existing simple electricity meters). For developing countries, long-term countracts and price forecasts are provided by utilities, but a large difference is that new industrial and commercial facilities are built knowing that, from the start, there will some day be spot pricing, basic designs incorporating controls, storage, fuel switching and enabling fast, large responses to be made to changing prices. With large labour forces available in developing countries, digital hardware is not needed for control, yielding more price-responsive industrial loads than in developed countries, with industries built assuming highly reliable, low prescribed price electricity. Developing countries' markets have faster, larger impacts on generation mixes and generation and transmission capacities than do developed countries' markets.

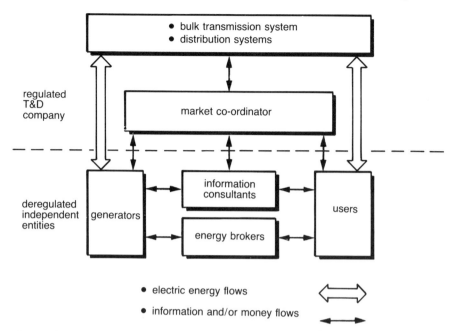

Figure 6.13 *A deregulated energy marketplace (Source: SCHWEPPE, F.C. et al, 'Spot pricing of electricity', 1988, Kluwer Press, Norwell, Massachusetts, USA, Figure 5.1. Reproduced by permission of Kluwer Academic)*

6.5 Regulatory aspects

6.5.1 Towards deregulation

Spot pricing of itself opens the door to some deregulation of generation, transmission, autogeneration, cogeneration and to wheeling. A basic deregulated generation market, shown in Figure 6.13, has three participants:

(1) A single, regulated utility controlling transmission and distribution, acting as middleman or marketmaker
(2) Many independent generators, cogenerators, autogenerators, etc. selling to the marketmaker amounts decided upon each instant
(3) Consumers buying from the marketmaker amounts decided upon each instant.

Table 6.3 shows the basic functions of the deregulated market of Figure 6.13.

6.5.2 Regulated marketmaker

The simplified single transmission and distribution marketmaker utility of Figure 6.13 acts as an intermediary physically and financially, buying all electricity offered for sale at each instant and selling all electricity consumers demand, periodically collecting all consumer payments and paying all generators for what they produce. The difference must cover the marketmaker's costs

176 *Pricing*

of expanding, maintaining and operating the sector. The marketmaker's duties include:

(a) Building, maintaining and operating the transmission and distribution networks
(b) Determining the spot price, communicating this to all independent generators and consumers
(c) Controlling system stability through pricing, demand managment and spot price-quantity contracts mentioned earlier
(d) Collecting monies from consumers and paying monies to generators.

The marketmaker must be at least financially regulated by the traditional rate of return on assets framework, financial targets, or something similar, as described in earlier chapters. Revenue recovery to cover capital costs for extensions can be done by modifying the spot prices in a logical pattern. Other allied aspects are covered in other chapters, e.g. under transmission planning in Chapter 7.

6.5.3 *Generators*

Generation represents the bulk of assets in most electricity sectors. In the deregulated market of Figure 6.13 these assets are owned by a number of generators which:

Table 6.3 *Basic functions of an electric power system (Source: SCHWEPPE, F.C. et al, 'Spot pricing of electricity', 1988, Kluwer Press, Norwell, Massachusetts, USA, Table 5.1. Reproduced by permission of Kluwer Academic)*

Short-term operation and control

  Dispatching generation minute by minute.

  Operating the transmission system during normal conditions; system security, system dynamics, relaying.

  Controlling the overall system during emergency state conditions such as sudden generation or transmission outages.

  Setting prices to customers; choosing which customers will receive electricity during system emergencies.

Long-term operation and planning

  Unit commitment, maintenance scheduling, and fuel purchasing.

  Investment planning. Choosing what kind, where, how large, and when to build:

  Generating units
  Transmission lines
  Local distribution systems.

  Forecasting future conditions.

Pricing 177

(a) Build, maintain, and operate generation and storage capacity
(b) Sell electricity to the marketmaker at the current spot price
(c) Meet environmental and other constraints, e.g. national conservation requirements
(d) Are not subject to direct, heavy regulation
(e) Are barred in the public interest from explicitly co-operating with other generators in their area or owning too much generation in one region
(f) Basically are profit motivated
(g) Act socially beneficially only because of competition, or government action.

Traditionally, generator ownership is geographically based, but under a deregulated market it will also be based on function, i.e. different generators specialising in different plant types (coal or nuclear steam combined cycle or peaking plant), able to specialise rather than maintaining across-the-board proficiency. Thus one utility owning one large power station (a case detrimental to both the national economy and to consumers) will be avoided.

6.5.4 Consumers and others

From a functional viewpoint, consumers in Figure 6.13 act independently of whether the market is deregulated or otherwise. As previously described, large industrial and commercial plus sophisticated residential consumers, will see 1-hour and 24-hour updated spot prices, small residential consumers seeing billing period updates. Yet the deregulated market must be stable to have large numbers of price–quantity transactions dealing in the short-term (seconds to minutes) required for emerging system control, e.g. for carrying demand-side spinning reserve. Although going from a regulated to a deregulated market has little, if any, direct institutional impact on consumers, an underlying assumption always in deregulating is that prices will come down in the long-term. It will normally be more meaningful in practice to disaggregate the single marketmaker into a single bulk transmission utility marketmaker and several local regulated independent distribution utilities. Figure 6.13 shows other participants: information consultants forecasting future spot prices, short, medium and long-term; and agents or brokers, arranging side letters and contracts to spread risks between all parties concerned.

6.5.5 Balancing supply–demand

Under deregulation, generators are not centrally dispatched as is traditionally the case under regulation, especially under public ownership. The marketmaker of Figure 6.13 sends each generator a spot price and each generator self-dispatches, usually producing some output if the spot price exceeds marginal running costs. The marketmaker keeps a supply–demand balance by continuously adjusting the spot price, when demand is minimal the spot price declining until the market is cleared efficiently, when only generators with low running costs remain, e.g. nuclear, some base-load steam, wind, etc. With increased demand, the spot price rises to where storage owners, who bought cheap electricity to store, start selling back. On the demand side, consumers reschedule electricity-intensive actions to low spot price times, responding to pricing so as to maximise net benefit. At times of a sudden system misfortune, e.g. a large generator outage, the spot price increases immediately. Generators already producing then increase their output to maximum; consumers' process control computers delay automatically the 'on' of air conditioners, furnaces, heaters, pumps, etc. Other generators start producing; reservoir hydro plants increase output. For a severe outage, the marketmaker uses some price–quantity interruptible loads, at least for the first few minutes so crucial to system stability. Marketmaker measures then gradually lower the spot price. If it still remains on the high side, then some consumers close down some operations to save funds, but equilibrium can only be restored at a higher price.

Under traditional system operation, central control knows, or can find out approximately enough, the actual operating costs of all generators, so that the marginal fuel cost part of spot prices can be easily computed as the short-run marginal cost, plus any quality premium to ensure supply–demand balance. Under deregulation, the regulated marketmaker does not probably know precise generator cost characteristics, especially from a few years after deregulation. The marginal operating cost part of spot prices can then be determined empirically by watching the price–demand response of individual suppliers at different times. The quality of supply premium is under the marketmaker's own expertise. Under the transparent information system described earlier, marketmakers would ask generators to confidentially furnish per unit operating costs to facilitate equitable system dispatch. In Figure 6.13 the marketmaker, knowing transmission or distribution details directly, adds the network components of the spot price. Under a transparent information system, therefore, spot price evaluation is almost the same for regulated or deregulated marketplaces. But it remains to be seen how long any system can contain transparency to this extent in a confidential world.

6.5.6 *Long-term planning*

Long-term planning (up to twenty years) aspects of Table 6.3, carried out in the deregulated conditions of Figure 6.13 still requires forecasts of critical parameters to be made by all utilities, especially fuel prices, cost of capital, electricity demand and market shares. Traditionally utilities have been expected to act in the best interests of the nation and of consumers by minimising total (capital plus operating) costs over time. Under deregulation maximising utility profit takes over automatically but, with efficient markets, this should also improve consumer welfare, depending upon the amount of real competition. Since ownership and investment decisions are both decentralised under deregulation, private utilities build plant if they believe it to be profitable, an approach followed in any industry that sells its product in a spot market, e.g. oil, metals, commodities, where the basic investment rule is: if net revenues, appropriately discounted and tax-adjusted, are expected to be appreciably larger than building costs, then invest; this at least resembles the optimal capacity expansion process of regulated or publicly owned generator utilities.

In a fully regulated or publicly owned situation, vertical integration, however loosely applied, co-ordinates generators and network expansion plans, and networks by-and-large are not normally over or under loaded. This need not be so under deregulation, which could be a serious weakness. When a deregulated generator is considering a new power station, it will pay the regulated network utility(ies) to carry out load flows, system stability tests, etc., the calculations to determine whether the new generation is likely to set unfavourable quality of supply components of the spot prices. If the latter is so, the regulated transmission or distribution utility(ies) might offer to build more lines and to charge the generator(s) for their use, each deregulated generator then paying for the amount of the new line(s) used. To do this effectively, the network utility(ies) must forecast alternative new generation programmes. Under deregulated markets, separate forecasts are needed by each utility, generator or network, or these can be purchased from information consultants, or agents. Generators then base their long-term planning on forecasts of future spot price patterns over the planning time horizon, in effect these forecasts replace traditional plant availability and demand forecasts. Investment decisions still need forecasts also of fuel costs, cost of capital, etc. Many different independent forecasts will be made instead of the single central forecasts made under fully regulated or publicly owned situations. This should have the many advantages of variety. Moreover, price feedback must make spot pricing patterns easier to forecast. There remains the subject of risk of uncertainty and who should bear this under deregulation. Generation has always been a risky business but, traditionally, the risk has been catered for by having expensive, large planning and operating margins of spare plant. Under deregulation the risks are borne by generators and consumers somewhat differently, although in total the risks may be no more. Long-term fixed-price, fixed-quantity contracts permit cogenerators, autogenerators, etc., to hedge some of the risks. Such contracts are like commodity futures contracts, purely financial, purchased by producers, consumers or speculators. Energy brokers in Figure 6.13 expedite such transactions by bringing buyers and sellers together. But such contracts do not improve spot pricing efficiency, being based on fixed-quantities at fixed prices.

6.6 Commentary summary

Traditional pricing is summarised in 6.1.7. In many respects dynamic, spot and real-time pricings are but an extension of demand management and time-of-use tariffs into actual time, i.e. rather than a time prescribed in the future from LMC. Dynamic or spot pricing uses actual short-run marginal costs (SMC) at any instant, plus a quality of supply premium to balance supply–demand. It is likely to be extended considerably in the 1990s increasing into most commercial and industrial tariffs, with residences gradually following suit. Short-run dispatching factors, generator competition and dynamic spot pricing will help maximise outputs from cogeneration, autogeneration and renewables. The emphasis in the 1990s towards private ownership and funding, demand management, improved efficiency, conservation, short-term finance rather than long-term economics, competition, deregulation, marketmakers, buying electricity forward, futures electricity markets, etc., all emphasise the short-term at the expense of the long-term, thus favouring dynamic and spot pricing, because it is based on conditions at the time of electricity usage, rather than pricing prescribed from a long-term development programme, and specified well before consumption. Many institutional electricity factors need little modification for dynamic pricing to work whilst others need considerable changes: using the regulated grid as the marketmaker, rather than the central dispatcher;

encouragement of agents or brokers to assist generator, grid, distributor utilities, and consumers to carry out their short and long-term planning, investment and operation; development of a new information system to replace merit order generation schedules; removal of statutory obligations to supply; changed role of government or heavy deregulation towards light regulation or deregulation.

6.7 References and further reading

1. MUNASINGHE, M. (1990), 'Electric power economics' (Butterworths) Chapters 3 and 4; also, Berrie, T.W. (1983) 'Power system economics' (Peter Peregrinus) Chapter 9; also Turvey, R. and Anderson, D. (1976) 'Electricity economics' (Johns Hopkins University Press)
2. BERRIE, T.W. and ANARI, M.A.M. (1986), 'Joint supply-demand optimisation', *Energy Policy*, December 1986 pp. 515–527; also Munasinghe, M. (1985) 'Energy policy and integrated pricing framework': paper given to the *International Symposium on Energy Sector Management and Modelling* (Imperial College, London)
3. MUNASINGHE, M. (1980) 'Integrated national energy planning for the developing countries', *Natural Resources Forum*, Vol. 4, pp. 359–373
4. UNITED NATIONS (1972) 'Electricity cost and tariffs: a general study', New York
5. BERRIE, T.W. (1985) 'Economics of variable pricing in the UK electricity sector': internal paper concerning the work of M.A.M. Anari at Department of Electrical Engineering, Imperial College, London
6. WORLD BANK, (1984) 'Guidelines for marginal cost analysis of power systems' Energy Department paper No. 18; also connected Staff Working Papers, World Bank
7. STEINER, P.O. (1957) 'Peak loads and efficient pricing' *Quarterly Journal of Economics*, November, pp. 43–49; also, Hirschleifer, J. (1958) 'Peak loads and efficient pricing: comment' *Quarterly Journal of Economics*, August, pp. 82–85; (some analyses in this chapter are modelled on this Hirschleifer article); Nelson, J.R. (ed.) (1964) 'Marginal cost pricing in practice' (Englewood Cliffs, NJ); also Meek, R.L. (1963) 'The green tariff in theory and practice', *Journal of Industrial Economics*, July and November; also Millard, R. (1971) 'Public expenditure economics' (McGraw Hill) Chapter 8
8. Central Electricity Generating Board, Finance Department (1972) 'Bulk supply tariff; history of its development'; also, Francony, N. (1982) 'Theory and practice of marginal cost pricing: the experience of Electricité de France; also Hons, A. (1981) 'An attempt at an international comparison of electricity rates, Central Electricity Generating Board', Report by the UK Monopoly and Mergers Commission, HC 315, HMSO
9. BERRIE, T.W. (1984) 'Implementation of advanced tariffs': paper to *Conference on the Economics of Electric Power Systems, Electricity Tariffs and Industrial Reaction* (University of New South Wales, Sydney)
10. SCHWEPPE, F.C. *et al.* (1988) 'Spot pricing of electricity' (Kluwer Press, Massachusetts, USA) (parts of this chapter owe much to what is in this book); also, Berrie, T.W. (1987) 'Improving marketplace efficiency in the non-oil energy sectors', *Energy Policy*, August, pp. 315–328; also, Berrie, T.W. and McGlade, D. (1991) 'Electricity planning in the 1990s, *Energy Policy*, April, pp. 192–211; also, Berrie, T.W. (1983) 'Power system economics' (Peter Peregrinus) section 9.5

Chapter 7
Development programmes

7.1 Introduction

By now, established tradition means large steam power stations supplying expanding grids and distribution networks, shutting down old stations near city centres, concentrating on a few thermally-efficient generators readily accessible to primary fuels and cooling water, designed to optimum technology. Renewables still remain only a likely future development, presently more expensive than burning fuels but hopeful for wind and solar, e.g. solar farms on barren tropical land. With solar cells' manufacturing cost reducing, break-even costs should be reached in the late 1990s, allowing a significant solar contribution, e.g. one US utility expects to supply over 2000 MW of firm generation from renewables by 1994, about 15% of requirements. In many places, wind is a valuable resource, but with unreliable sunshine needs expensive control devices, making for difficulties in competing with conventional generation. Tidal power has construction and operation problems dealing with low head and large water volumes, as well as storage for firm power. However, the 1990s will see a change to smaller, non-steam power stations built much closer to load centres. Transmission and distribution will also see changes, e.g. lighter, compact, solid-state, more easily controlled switches with no moving parts, low loss transformers, worthwhile on energy saved. High voltage equipment will be encapsulated in resin or polymers, insulated with non-inflammable, non-toxic liquids. Modern control-cum-communication techniques will improve reliability, microprocessor-based protection and controls replacing electromechanical relays.

Conservation is only one way of consumers cutting costs. Using 'wheeling' a consumer not satisfied with electricity from one utility can deal with an agent or broker, for supplies from another source, especially applicable to transmission and large users (e.g. smelting or chemical industries depending upon cheap, efficient electricity to compete in world markets) but, increasingly, using distribution wheeling for smaller consumers. More efficient electricity use leads to all-round savings, also greater user comfort and convenience. Nearly half of electricity usage in lighting, refrigeration, house and office heating, air conditioning, etc., is probably savable by modern technologies and insulation; high frequency, electronic ballast fluorescent lights provide as much light as any incandescent lamp nearly five times their rating, or twice that of conventional fluorescent tubes with choke ballasts. Freezers and refrigerators are five times more efficient with thermal insulation and improved compressors. Heat loss from buildings is halved by loft and wall insulation, double glazing, etc., and air conditioned buildings are made almost self-sufficient by recycling air through

182   *Development programmes*

heat exchangers. Also, control by solid state controllers and sensors of lighting and heating substantially reduces electricity requirements. Modern processes require less electricity per unit of output than a decade ago, in particular production of plastic to replace steel showing considerable energy saving, and all modern electronic appliances such as television and radio sets take only a fraction of previous input. There are many more examples and these trends will continue into the 1990s and 2000s.

7.2 Generation programmes

7.2.1 *Plant types*

The three primary energy sources are [1] chemical, solar and nuclear; the three secondary energy sources, thermal, electric and mechanical, with many ways from primary to secondary, each path losing energy on the way. Two other important factors are energy storage and energy conversion. It is preferable to use primary energy directly rather than go through secondary, or to convert from one type to another. There is presently no economic direct route between the three primary energies and secondary electricity, only through heat, mechanical energy, etc. Specialist descriptions of generation types now briefly discussed are given fully in the literature [2].

Hydro plant converts solar primary into secondary electricity, the water being used also for irrigation, recreation, navigation, etc. Below heads of 70 m, hydro is classed as low head, e.g. run of river and microhydro. All hydro gets publicity and needs environmental maintenance. Lead times are five years plus; economic sites throughout the world get rarer. If hydro includes tidal energy converted to electricity, the high capital cost is locked into a fixed 50-year situation during which the socio-economic climate will change. Although hydro incurs no fuel costs, the clearing of silted up reservoirs may constitute an appreciable recurrent cost. The thermodynamic Carnot cycle [3] limits thermal efficiencies of steam generators to a little under 40%, with super-critical boilers adding a few percentage points, quickly tapering off, with super-critical boilers not widely used because of mechanical stress. Steam generators seem at their ultimate technology. Because electricity and gas are competitive 'prime' fuels, there is still often reluctance to burn gas under electricity boilers, even when gas is present in large volume, cheaply, environmentally friendly, and locally. Diesel generation is normally associated with small utilities, industrial works and developing countries, it being small, flexible, portable and modular. Slow speed (200 RPM) diesels' thermal efficiencies of up to 50% seem possible. Although diesels need more maintenance than steam generators, waste heat can be used to drive a small steam generator, the plant becoming a combined cycle plant. Diesels cannot efficiently become greater than about 20MW and there are problems with their parallel operation and with the grid. Combined cycle can add 20MW of steam capacity. Diesels have small lead times, are often obtainable virtually off the shelf, and their capital cost is low.

Gas turbines range from small 30MW/50MW to large 200MW. They can easily run in parallel and have good thermal efficiencies. Efficiencies and size will increase throughout the 1990s with new materials being used. They are quick starters–stoppers for demand management, have short lead times like diesels but perhaps not quite as short. However, they need more cladding than anything but slowspeed diesels. They are more suitable, if anything, than diesels for using their waste heat in a 30MW (plus) steam generator, forming a combined cycle. All combined cycle plants have higher thermal efficiencies than steam plants and with regeneration, efficiencies up to 45% plus. Most gas turbine problems have been rectified and stations of 300MW upwards are available for the 1990s, up to 2000 MW stations. Diesels or gas turbines are built first and the expensive steam stage committed only with load growth, but even then the diesels or gas turbines can still be run on their own when demand patterns make this economic.

Nuclear generators have never been considered economic in sizes less than 600MW, 1200MW being preferable. In any case, the stigma against nuclear in the 1990s will be:

(a) High capital costs, especially when all costs are included—loading, safety and decommissioning
(b) Nuclear pollution dangers
(c) High costs to ensure safety

Commercial geothermal [4] energy is real but with problems needing solving before using the full potential, some site-specific and some general, especially finding the correct balance between electricity and heat application for higher temperatures. At the other end of the scale, the use of heat pumps in association with lower grade geothermal is less economic than might earlier have been expected. Economic wave and wind generation varies between countries and geography, low to medium amounts being justifiable in places but not as successors to conventional generation [5]. Looking ahead, thermal voltaics are expensive, only justified for small, remote, electricity sources, e.g. for telephone exchanges. Despite increasing efficiency, problems remain concerning building banks of voltaics at reasonable cost, although these will decrease with numbers manufactured. Although needing maintenance their life is long. Fuel cells are becoming viable but still have electrode corrosion problems. With high cost, likely success is for outputs up to 10MW, used for urban peak lopping. Rechargeable batteries are most useful for auxiliary supplies in substations, for backup and urban peak lopping. Nuclear fusion [6] could still prove to be the 21st century breakthrough replacing conventional fuels, harnessing enormous energies in the light elements.

7.2.2 National electricity

All generator utilities, cogenerators and autogenerators must make at least outline surveys of national generation by size, type, fuel, thermal efficiency, age, availability, existing generators plus those under construction and/or committed, optimum load factors of all plant, their maintenance requirements, outage time and ownership. Large privately or publicly owned utilities will carry out the survey themselves, smaller utilities employing agents, especially in fragmented electricity sectors. Each such plant register must be kept continuously updated, especially when making investment decisions, when detailed attention

7.2.3 Wholesale market

To an extent which is warranted, everyone concerned must forecast the national wholesale electricity supply–demand position, allowing for competition, demand management, improved efficiency, conservation, etc., small outfits employing agents, examining medium and long-term load factors with regional variations, annual maximum demands (kW) and kWh, with demand profiles per hour/day/week/month/season/year; and for reactive power (kvar) as well as for kWs. Similarly, all concerned must forecast market shares, an iterative process with smaller outfits employing agents, leading to supply–demand scenarios for the national wholesale market in detail for each utility's cogenerator's and autogenerator's share, with confidence probabilities for the short, medium and long-term, at first ignoring transmission and distribution, i.e. until the number of scenarios has been cut down, when these can be added to find the joint generation, transmission and distribution optimum.

7.2.4 Optimum programmes

Traditional methods for determining least-cost generating programmes [7] assume a single node or an infinite grid, use total annual capital plus annual whole-system operating costs in merit order, the annual costs being present valued at a suitable interest rate and summed up to the planning horizon. The optimum programme is that with the lowest summed present value of cost, each alternative programme meeting the forecasted patterns of demand at the same degree of risk. In the 1970s and early 1980s, using computers, these optimisations and the prescribed pricing derived from them often became very complex, with large numbers of sensitivity analyses testing parameter uncertainty effects (Figure 7.1).

From the national economic viewpoint, fuel costs must reflect opportunity-costs (i.e. border or international or import prices) because fuel used domestically either reduces export earnings or increases import bills. Such fuels should also be given an economic rent for being non-renewable.

In figure 7.2 points A and B show present-day costs of domestic fuel and thermal equivalency prices of internationally traded fuels, respectively. With infinite reserves, fuel opportunity costs are equal to supply costs along AC. With a good export market, or substitution possibilities for imported fuel, the opportunity cost follows path AD. With early depletion of domestic reserves expected, the price path is AFE or IJE. JL is the rent at depletion time T. If T tends to infinity, IJ tends towards AC, meaning if reserves are large, opportunity cost is the supply cost. If T tends to zero, IJ tends towards AD.

During the late 1980s [8] electricity sector decision-makers worldwide drew attention to serious defects in the traditional methods of planning, as listed in Box 7.1.

The axioms of planning for the 1990s are given in Figure 7.3 and in Box 7.2 and Box 7.3.

Development programmes 185

stage i	ii	iii	iv	v	vi	vii	viii	ix
underlying market	supply-able market	alternative planting type and programme and site			least cost alternative	high priority use of country's resources ?	subsidy required ?	delay or accelerate ?

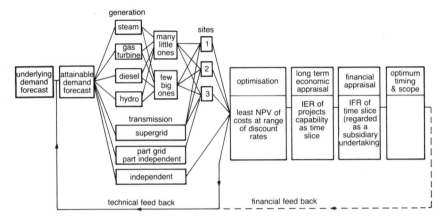

Figure 7.1 *Stages of economic-financial assessment of projects or plans*

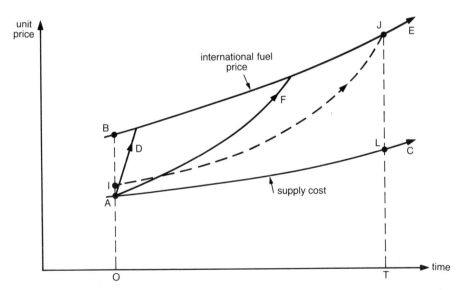

Figure 7.2 *Dynamic price paths for domestic exhaustible energy (Source: THAKOR, V. 'Economic evaluation methods of power projects', paper to IEEE Winter Meeting Feb 2, 1983, New York. Reproduced by permission of the IEEE)*

Box 7.1 *Defects in traditional method for electricity planning*

- Failure to forecast accurately loads, costs, construction times, plant availabilities, plant efficiencies, etc., even one year in advance
- Serious over and under planting, these often running in cycles
- Excessive utility debt service requirements
- Failure of prices prescribed from the optimum least-cost development programme to give the right signals and plant–load balances in the event
- Wrong optimum generation plant mixes, by fuels, sizes, type, etc.
- Inability to withstand sudden changes, let alone catastrophes
- Least-cost solutions not being so in practice
- Divorce of long-term planning from ownership and accountability
- Inability to cater properly for autogeneration and cogeneration
- Almost complete exclusion of consumer reactions from the planning process
- Absence of the effects of planning other fuel sectors on electricity planning
- Selling improperly computer softwares for planning, leading to loss of faith by decision-makers in computer-derived programmes
- Inability to cope with uncertainty, despite its elaborate, complicated sensitivity/risks analysis
- Failure to support common sense cases for retaining indigenous fuel resources
- Preventing electricity outputs from artifically boosting the economics of irrigation in multi-purpose schemes
- Failure to treat electricity as a multi-product
- Inability to be used for entrepreneurial type planning
- Failure to make optimal use of the private sector for electricity supply

7.2.5 Finance

Alongside finding least-cost long-term development programmes and prescribing prices from these, traditionally there have been ways of dealing with the short-term, i.e. ordering of actual plant and producing a financing plan. To do this, the optimum long-term development programme is given detail in the early years to determine the plant ordering points, including lead times. Final decisions on short lead time plant (diesels and gas turbines) are put off until six to eighteen months before requirement, but other plant is ordered much more in advance, especially nuclear and hydro, because of their environmental effects. This leads for the short-term to:

(a) A utility financing plan showing all sources of finance, internal and external, debt service requirements and required financial engineering, i.e. the fitting of the types of finance to the different components of the project
(b) A pro-forma set of forecasts of utility annual financial statements, e.g. balance sheet, profit and loss account, sources and application of funds, etc., for the next (say) five years.

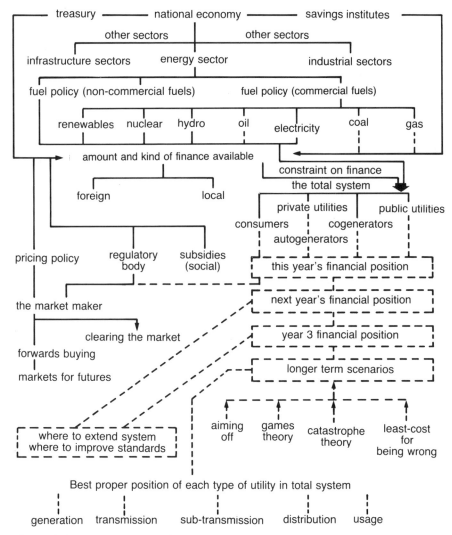

Figure 7.3 *The new approach to power sector planning*

There are many ways of modifying traditional prescribed tariffs to ensure correct revenue recovery, all well covered in the literature [9], and this is basically an accounting question. The method preferred depends upon:

(a) Ownership; shareholders' requirements are different from local or central governments'
(b) Legalities; laws governing utilities vary considerably
(c) Payback; private ownership and funding demands shorter paybacks periods than public ownership and funding
(d) Returns; private ownership requires larger annual returns on turnover or assets, and also larger returns on capital than public.

Box 7.2a *Axioms of electricity planning for the 1990s*

- Choose the development programme least likely to be wrong in the event
- Build into development programmes a degree of flexibility
- Choose development programmes capable of dealing with sudden changes, even catastrophes, without collapsing
- Adopt scenario planning, rather than deterministic planning
- Treat decisions as individual moments in a longer process, itself within the wider backdrop of national, sector, and utility planning
- Bring the demand side, including consumer behaviour, fully within the planning process
- Use interactive, computerised, graphic display, computer models of planning alternatives for use by decision-makers
- Treat planning as a product of the market and pricing, not vice versa
- Use the theory of games, war-games, catastrophe theory in electricity planning
- Obtain multiple-objective solutions rather than single least-cost solutions
- Deal also with utility capital and running costs in financial terms, not only in national economic terms as in the past

Box 7.2b *Pricing and planning in the 1990s*

- Count *DM projects and conservation* methods as alternatives to new projects
- Count *annual returns on assets* as the meaningful guide for the utility
- Governments only to be concerned with *discount rates and economic returns on investments*, utility only indirectly concerned
- *Price* to approach towards perfect competition in electricity markets
- Use *short-run* not long-run marginal costs
- Assume *dynamic, real-time pricing* gradually takes over from prescribed pricing
- Take proper notice of the economics of *cogenerators, autogenerators*, etc.
- Discontinue using *high discount rates*, even for electricity systems in developing countries in high growth, because for financial purposes all years, even year 30, can be thought of as having near equal weight with the first year

Box 7.2c *Treating basic parameters in planning in the 1990s*

> - Treat the *reliability standard* as part of the joint supply–demand electricity system
> - Treat *economies of scale* as suspect on capital, but as real on fuel
> - Measure justification for *advance, automatic system control* against short-run costs; this not being a long-run exercise, which seldom makes a good case for this equipment

7.2.6 Commentary summary

Questions to be addressed with respect to traditional generator planning as it should be used for the 1990s are given in Boxes 7.3a and 7.3b.

7.3 Transmission programmes

7.3.1 Formulation

The traditional formulation of transmission programmes is well established and documented [10]. Traditionally, transmission (and distribution) planning resembles generation planning in:

(a) Finding, using approximately the same methodology, optimum least-cost long-term (generation plus) transmission and distribution programmes, accompanied by an analysis to check how the optimum solution is

Box 7.3a *Questions with respect to traditional generation for the 1990s*

> - *Electricity demand.* In so far as it concerns the programme/utility/project under examination: is the information sufficient? What are recent trends in kWh, daily, weekly, annually, and annual peak loads (kW)? What are consumer classes' break-down, plus demand patterns, plus future trends? Is there any unsatisfied demand? Which forecasts are important, short, medium or long?
> - *Pricing and demand.* What is recent history on tariffs, how near are tariffs to the economic efficiency levels? What are the rules for adjusting tariffs by government, regulatory authority, utility? What about revenue recovery, financial obligations, annual rate on assets? Financial statements, past and next two or three years? Future trends based on prescribed and dynamic pricing as alternatives? Are these properly allowed for?
> - What is the *existing generation* (possibly also transmission and distribution), primary fuels, losses, overall efficiency, committed plant, what are the capacities of generators, plant types, fuel types, operating margins, plant-load balances?

Box 7.3b *Questions with respect to traditional utility planning in the 1990s*

- With respect to *electricity institutions*, e.g. generator utilities: how much autonomy? What is their format? Under what regulatory authority, and what are the utility's financial obligations, to whom? Under what format? What are relations like between utilities, generator, transmission or distribution, or with regulating authorities, or with government or with financing sources? What is each utility's format, statutory authority? Is it a public utility, private utility, municipal corporation? What is the size, annual turnover, staff numbers, skilled, unskilled? What is its investment profile, electrical energy sold and annual maximum demand?
- What is the likely need *for future plant*, from growth of peak demand and annual energy, plant scrapping schedule, risk standard, planning plant margin? What plant capacity is required with various plant types and sizes and timing of new plant?
- What is the *least cost, long-term development programme*? What discount rates should be used for capital costs, fuel availabilities and prices? How much is the least-cost programme sensitive to changes in parameters? What are the scenarios for proceeding from existing plant and existing prices?
- What are the utility's immediate *revenue requirements*? What are the consequences of these requirements on tariffs, financial viability of the utility? How will the utility finance the future? By borrowing? What are the pro-forma financial statements for the next (say) five years?
- What are the *new factors* for the 1990s? What are the likely effects of privatisation, regulatory bodies, market competition, demand management, improved efficiency, conservation and environmental maintenance requirements?

sensitive to uncertainty in the value of the parameters and including an economic analysis of reliability standards

(b) Finding from the optimum transmission and distribution programme how the costs of this programme should be reflected in electricity pricing.

Electricity is transmitted wholesale from generators to demand areas and the network structure must adjust to the national and regional energy supply–demand patterns. Electricity is today produced and distributed at high voltage, 3 kV to 33 kV, possibly 66 kV for distribution, and transmitted at even higher voltages, 110 kV to 500 kV, possibly higher, by both alternating or direct current. Transmission at such high voltages has advantages in reduced losses and economising in overhead conductors, but calls for high towers, costly insulation and extra outlay on extensive, ugly, transforming-switches stations. Transmission costs roughly increase with distance and a balance must be struck between:

(a) Economies for a demand centre from using a cheaper energy source than local energy, the remote energy transmittable as electricity, gas, oil, or coal
(b) The cost, financial, economic and environmental, of getting that remote energy to the local demand centre.

One of the basic economies obtainable from linking generators is the pooling of generation standby requirements, required for unforeseen breakdowns; such grid interconnections can save 30% to 60% compared with otherwise separate systems pooling. But underground can be twenty times as expensive as overhead transmission, justifiable for either short sections or special circumstances (e.g. underwater cables), for overriding safety and for environmental maintenance.

7.3.2 Options

Flexibility is required [11] just as much for transmission as for generation planning because electricity is but one form of transportable energy. In developing countries, transmission often starts from a 'green field' situation, but most transmission options are concerned with extensions, or the overlaying of a higher voltage network onto an existing network. Voltage levels are fully standardised at least by world region. Before deciding on extra transmission, load-flow studies are needed to ensure the expanded transmission network will be properly loaded under most conditions. Transmission limits depend on maximum thermal loading, also steady-state and transient stability limits, instability today being almost automatically predictable by network 'state analysis'.

Electrostatic fields round transmission conductors produce electrical stress, noticeable in sparks and corona, giving interference with radio, telephone and possibly humans; the remedy is to increase conductor numbers. Important for transmission in the 1990s are debates over environmental maintenance and keeping the ground electrostatic gradient to (say) 10 kV per meter. In the 1990s options to increase transmission throughput per circuit are limited with one exception, namely, moving from 3 phase to 6 or 12 phase, but this would threaten possible worse problems with high open circuit voltages when a line is switched out routinely or in an emergency. Open circuit overvoltages can be just as much a limitation on transmission as lightning, switching or other surges.

With respect to safety options on transmission circuits, autoreclosure after faults is usually a good idea, but perhaps only single phase. In the 1990s it is likely that more direct current (DC) transmission will be both constructed and planned than ever, especially in North America. DC transmission can interconnect two systems using different frequencies and it does not increase the short circuit dangers. However, it still cannot be directly switched, a serious disadvantage.

7.3.3 Rules

When deciding the best network structure the various major rules for transmission are given in Box 7.4 in the form of relevant questions to be asked.

The four traditional functions of transmission are given in Box 7.5. All transmission functions are interrelated, the weight given to each changing over time, e.g. due to technical advances and greater emphasis given to environmental considerations, particularly for nuclear power.

For the 1990s further interconnections in developed countries will run into difficulties in:

192   *Development programmes*

(a) Technical design; handling high loads over long distances necessitates very high voltages, heavy conductors and AC to DC conversion plant, if DC transmission is used
(b) Operations; maintaining stability under fault conditions and ensuring fault level limits are not exceeded, requires most sophisticated central systems techniques
(c) Installation; transmission cannot be installed in small packages as and when required, and its full potential may not be utilised for several years if at all, leading initially to over-capacity and higher than strictly necessary capital expenditure
(d) Politics; international problems often frustrate fully using interconnection.

7.3.4 *Wheeling*

Wheeling is basically using the electricity network as a 'common carrier' [12]. Ownership of the common carrier basically affects wheeling, but in a manner not yet fully understood. Wheeling was first devised for wholesale electricity at voltages of between 110 kV and 220 kV, not too high for some consumer connections to be taken directly from the grid. Experience indicates this as the ideal wheeling voltage. However, it is suggested that:

(a) Wheeling at above 220 kV is desirable, even necessary, to make wholesale dynamic or spot pricing for large industrial consumers work properly
(b) Wheeling at lower transmission voltages, even at primary distribution voltages, 33 kV, 11 kV, will probably in the end be necessary for dynamic or spot pricing of small industrial and all commercial loads
(c) Carrying out wheeling for retail loads, e.g. at 11 kV, 3 kV and 415/220 volts at present seems difficult and is possibly not desirable because of losses.

There are basically four main types and six main forms of wheeling, as given in Boxes 7.6 and 7.7, respectively. Electricity losses vary with the square of the loading. If system configuration and parameters are known, losses due to

Box 7.4 *Major rules for transmission: relevant questions*

- Should *generators* be interconnected regardless of load centres?
- Should *load centres* be interconnected regardless of generators?
- Should *generators and load centres* be interconnected?
- How many *tiers of voltage* should there be?
- How much *interconnection* should there be under each case and tier above?
- Is a *'grid' structure* too overall costly, despite it giving greater security and ease of connecting new generators and wholesale loads?
- *Extra questions for the 1990s*: What transmission structure is favoured by privately owned competing generators? Is it significant that, in North America with traditionally private utilities, a tie-line structure exists, but with heavy regulation at local, state and federal level? Is the transmission network structure affected by the grid being used for wheeling, as a common carrier, as a marketmaker, as a spot price setter?

Box 7.5 *Traditional functions of transmission*

- *Planned transfer:* energy sources are in one place and energy sinks are in other places, e.g. electricity is exported from Switzerland to Germany; for hydro no opportunity exists to transfer energy to load centres without transmission, but for gas, oil, and coal, transporting the fuel must be compared with electricity transmission; a trade-off must similarly be made between generating at the fuel source and generating at the load centre, depending on cooling water and space available, also fuel transporting problems
- *Pooling reserves:* transfers for pooling reserves differ from planned transfers in that they cannot be fully evaluated before interconnection is actually implemented, because the timing, duration and magnitude of the actual transfers is largely unknown until maintenance or a breakdown occurs. The pooling of generation reserves by transmission enables both capital and operating costs to be saved; generation types diversity can also be increased, enabling all generation types to be operated to their optimum output
- *Economic operation;* the whole generation–transmission system should be made to operate as one economic unit by grid dispatchers or grid marketmakers for the wholesale market, taking full advantage of the complete range of generation operating costs; without interconnection the system would be fragmented into numbers of smaller systems which would not have as many individual steps of incremental operating costs to employ for least-cost operation at any instant
- *Siting of generators and load centres;* with a grid, generators are easily sited to fully suit the location of fuel, cooling water, amenities etc., as can new load centres, industrial zones being easily located fully to suit raw materials, population, fuels, transport, etc.

Box 7.6 *Four main types of wheeling*

- *Utility to utility*, e.g. before 1990 for the UK, between England and Scotland, and between England and France
- *Utility to private consumer*, e.g. by contracts between utilities and private consumers
- *Private generator to utility*
- *Cogenerator* or *autogenerator* to *private consumer*

wheeling can be worked out, thus the cost of losses, and then apportionment of losses. If part of the grid reaches performance limits due to wheeling there will be a wheeling cost which can be allocated for this, which could be annuitised as a function of individual transmission line loadings or of total system demand, and the annuitised costs treated as operating costs, e.g. as in peak demand pricing in most bulk tariffs, worked out after the peak is known.

To determine wheeling tariffs it is necessary to work in MW-miles of wheeling, or find the structure and level of use of the network under wheeling, preferably by load flows. Perhaps there should be no distance charge, just entrance and exit charges, no MW-miles charges, just flow level charges. There could be spot price buying and selling at any instant, differences between buying and selling prices being the cost of wheeling, on the basis given in Box 7.8.

7.3.5 Regulation

Minimal regulation of grid utilities occurs if contracts are set up directly between generators and distributors; but the grid can operate as a common carrier, or buying and selling marketmaker for wholesale electricity; in both cases the regulation is similar. The regulator could simply monitor the charging of the generator incremental cost in merit order and of the transmission facilities. Since the grid forms a natural monopoly, the basic price control rule is of the type retail price increase (RPI) minus a factor 'X' met with in Chapter 6, and choosing X according to the annual rate of return on assets or turnover, this being the easiest available information. This formula has the advantage to consumers of exerting downward pressures on costs as their allocation in any natural monopoly is somewhat arbitrary and could otherwise lead to excess charges being made to consumers. When the grid utility acts as purchasing agent, additional regulatory problems occur in the division of monies earned as a consequence of the difference between the grid's buying and selling prices, i.e. its allowable profit, some of which is needed to help finance transmission extensions. Maximum regulation of grid utilities occurs if electricity buying and selling contracts are set up through the grid. Thus, as a wholesale marketmaker the grid utility can buy electricity from competing generators and sell it on to distributors, becoming the electricity pricing centre. It then requires price

Box 7.7 *Six main forms of wheeling*

- *Mandatory*, e.g. forced, negotiated prices, controlled by regulatory bodies or by published tariffs
- *Money*, e.g. generators run in merit order with wheeling costs worked out afterwards, difficult but possible to use with ex-post pricing worked out from preconceived electricity package profiles
- *Contract*, e.g. money wheeling under a specific contract of the rights to supply or buy electricity; or electricity supplied under arbitrage without contract, because of difficulties of proving contracts in practice
- *Energy* e.g. suppliers agree to start up generators onto the grid from (say) 7 am to 5 pm of 100 MW, while consumers agree to load the grid to 100 MW for same period
- *Marketmaker*, e.g. grid utility, acts as marketmaker for all supplies and consumption
- *Regulatory*, e.g. in unregulated, competitive marketplaces, sometimes similar to market maker wheeling, sometimes resembling the single generator and area distribution utilities situations, e.g. in UK up to 1990

Box 7.8 *Bases for calculating losses in wheeling*

- *Empty system*, i.e. system assumed able to operate under all loads; no other loading except wheeling loading
- *Vector difference*; i.e. wheeling loading assumed superimposed vectorially on normal loading as in the last case, causing maximum losses, but system changes may be negative (see below)
- *Positive difference*; i.e. assuming only increasing line flows are due to wheeling
- *Net effect*; i.e. assuming increasing and decreasing changes are due to wheeling
- *Critical path*; i.e. path for wheeling through network specifically defined
- *Marginal cost allocation*; i.e. changes in system losses and generation redistribution caused by system constraints due to wheeling noted, revenue reconciliation techniques operating, plus capital recovery costs, used to form wheeling prices

regulation according to a more complex formula, $(RPI - X + Y)$ where Y is a mark-up factor. Purchasing tariffs will in the 1990s need to be restructured to eliminate anti-competitive problems of most existing bulk supply tariffs.

To ensure cost minimisation of supply security, penalties must be levied if the grid utility strays from generator merit order for electricity purchasing, in so far as it knows the merit order, or if generators fail to provide contracted supplies, the best regulatory deterrent framework lying somewhere between these two extremes. The regulators' duties are to: consider consumers' and shareholders' interest; ensure supply security; support government energy policies; and create a competitive, efficient electricity sector and electricity marketplaces.

7.3.6 Commentary summary

Traditional techniques for selecting the optimum, least-cost transmission development, as developed in the 1970s, will not change much in the 1990s, and new technologies in transmission are unlikely. The transmission 'grid' function will decrease as generator siting and mixes, utility ownership patterns, and pricing mechanisms change, especially changes towards flexibility. One important new grid function is as wholesale marketmaker, paralleling generator deregulation and wheeling, the latter in various forms. A sequence of changes is necessary, starting with the 1980s, worldwide conventional privately or publicly owned, regulated generation plus transmission utilities. The goal of the 1990s is efficient deregulated, competitive generation, but regulated transmission and distribution, applied locally, regionally and nationally. The steps will be dictated by the changes in electricity pricing from prescribed to dynamic pricing, each of the steps being mirrored in changes in the functions of transmission networks:

(a)   First, transfer to dynamic from prescribed pricing, starting with buyback from autogenerators and cogenerators, legislation often mandating such buybacks but being vague about prices

- (b) Then announce a schedule for subsequent steps, allowing consumers, vendors, transmission utilities, and potential generators sufficient lead times
- (c) Allow all consumers to pay the dynamic prices
- (d) Make wheeling mandatory over the transmission network
- (e) Separate all management and operation of generation from those of transmission (distribution)
- (f) Sell distribution networks to municipalities, local authorities, power boards, etc., if they want them (see next section)
- (g) Divest any generation assets from combined generation and transmission assets
- (h) Make transmission subject to regulation and adapt the grid utility, the central dispatcher of national utilities, as wholesale marketmakers with wheeling

7.4 Distribution programmes

7.4.1 Types

The retail part of electricity has not changed much in the last half century [13]. Distribution is not affected by the recent changes in emphasis mentioned earlier, except with respect to privatisation, cogeneration and autogeneration. What is at last becoming better documented are the principles behind distribution planning, on the lines of least-cost programmes as for generation-transmission planning, which can be logically developed for distribution. However, distribution planning principles are easiest to illustrate from the concept of a building up from new small systems, either as extensions to existing well established systems, or as 'green field' developments. The first planning choice is whether supplies will be from the well established (grid) system or not. In most cases the grid will provide the cheapest, reliable supply if the new demands are within economic reach of the grid. Basic, non-grid generation is most commonly diesel or gas turbine generators and there are millions of these in use globally. Petrol generators are widely used to produce small amounts in developing countries [14].

Small hydro has a long traditional use in some rural areas and could be used in many more. Wood-fired boilers to produce process heat can generate electricity in sawmills and other rural industries; sugar factories use their waste bagasse similarly. Windmills are also used for small-scale generation, also photovoltaic cells. In grid-based distribution programmes the utility gradually expands the area of supply around existing demand centres on the grid. As the grid is extended further, areas come within reach, the process accelerating until the whole area is covered, and the extensions become part of the established system. In large areas full grid coverage may neither be possible nor economic until regional grids are established, these grids being first used for distribution. In the build up from a small new extension, the pattern of distribution lines is determined by the number and dispersal between potential consumers. Where clustered in villages, simple extensions from the grid may be economic, the main attraction of grid-based distribution again being reliability, versatility and often cheap electricity, compared with isolated diesel/gas turbine systems. Grid maintenance and management costs are low per kWh, wires, poles and transformers needing little attention, whereas isolated generators need constant attention and regular supplies of fuel and spare parts, both difficult to come by in remoter areas. Grid schemes are also the most flexible, accommodating large load increases without major modification, capacity increases usually being providable relatively cheaply, particularly as provision was probably made for them when the system was first designed.

In developing countries, however, many distribution systems are heavily overloaded and subject to frequent breakdowns. In the longer term, following improvements and extensions of the grid, the benefits of an improved supply will flow to all, existing and new consumers alike. It is impossible to give meaningful cost figures for grid extension schemes on a general basis, distribution projects varying in types with number of step-down substations needed, distance from the grid, type of area and loading, and how dispersed are consumers. Increasing the number of consumers supplied brings a proportional reduction in cost per connection because there is usually adequate basic capacity allowed for in the basic design to meet likely demands. The running cost of grid extensions are in repair and maintenance. The lives of poles, wires, etc., depend environmentally on conditions and equipment quality, but are normally expected to be about 20 to 30 years. Table 7.1 gives costs per kWh for a range of small to medium size diesels. For distribution planning, decisions about diesels must be based upon actual installed costs, actual loads and working arrangement, and the local prices for fuel, spare parts, wage rates, and other operating requirements.

Micro hydro refers to plant below 100 kW, mini hydro to 100–500 kW, and small hydro to 500–600 kW. Frequently, small hydro is a general description for plant below a few MW. To produce electricity such small hydro uses a turbine, not a waterwheel, the output depending on quantity of water and head through which the water falls. The efficiency can be over 90%. All such hydro needs to be carefully designed and built to work properly and to survive the damage caused by floods occurring (say) once every five or ten years with 20 times more water than in normal wet weather flows. If the hydro is distanced from consumers more than 2 km, a transformer is used to step up the voltage to avoid excessive transmission losses, a step-down transformer being then required at the other end. In addition, a variety of control and safety equipment

Table 7.1 *Calculated costs per kWh for small to medium diesel systems (Source: FOLEY, G. 'Electricity for rural people', (1990), Panos Rural Electrification Programme, 1990, The Panos Institute, London. Reproduced by permission from the Institute)*

Output in kW	10·00	20·00	30·00	100·00
Capital cost ($/kW)	700·00	500·00	400·00	250·00
Working life (hours)	7000·00	10000·00	12000·00	25000·00
Fuel consumption (l/kWh)	0·53	0·42	0·35	0·19
Costs (US cents/kWh)	41·00	28·00	19·00	12·00

needs installing, providing protection against lightning, system surges or other damaging incidents. The number of potential, small hydro sites worldwide is large, particularly in mountainous areas with high rainfall. Yet the construction and operational difficulties should not be underestimated, small hydro being a complex installation. The potential for community–run distribution using small hydros is attractive and is likely to be used more widely in the future, due to distribution systems being often privately owned and remaining separate from the grid for a longer period than in the past.

Focusing unduly on technology, it is easy to forget the substantial administrative needs of distribution systems. If development programmes are to succeed, the first requirement is a means of identifying and registering consumers and likely future consumers; then arrangements for legal agreements, covering rights and obligations; then provision for inspection and approval of consumer connections; also provision for repair services. There must be arrangements for billing and collecting from consumers for connection fees and electricity consumed. A variety of methods are used to carry out these administrative tasks, some devolved to the local community, others carried out by the utility or outside agencies. When distribution is based on grid extensions, the necessary administrative systems and staff are already available in the utility, if the same utility owns the grid, but not otherwise.

7.4.2 Reliability

Table 7.2 illustrates that a large proportion of total outages is at the distribution level although few consumers are sometimes affected. It is difficult in distribution planning to specify the optimum combination of scheme plus reliability standard, in the same manner as in generation or transmission planning, (Chapter 3), each utility making its own rules, usually working from the proven reliability standards of individual system components, singly and in combination:

(a) Transmission step-down substations; i.e. switchgear, transformers, instruments, etc., although this in practice may be a transmission, not a distribution responsibility
(b) Primary distribution; i.e. 66 kV, 33 kV, 22 kV, 11 kV, 6 kV, overhead and underground lines
(c) Primary distribution step-down substations; switchgear, transformers, instruments
(d) Secondary distribution; 3 kV, 600 volts, 400/230 volts, 220/110 overhead and underground lines
(e) Consumers' connections; at primary, secondary, distribution levels.

7.4.3 Distribution markets

Although some wholesale loads will always be supplied from distribution, most demands are retail. Each distribution utility must assess:

(a) Present total size, scope, composition, etc., of national retail market
(b) Present utility share of national retail market
(c) Future size, scope, composition, etc., of national retail market, five years in detail, a further ten years in outline

Table 7.2 *Percentage of forced interruptions by voltage*

Voltage of distribution system	Number of consumers affected (at most)				
	10	100	1000	10 000	100 000
	Percentage of Total Interruptions				
132 000 and above	5	8	14	50	100
33 000	15	18	28	40	0
11 000	5	26	46	10	0
200–600	35	48	12	0	0
Others	40	0	0	0	0
Total	100	100	100	100	100

	Percentage of faults restored by time quoted by voltage and type of distribution system			
	Outage time in hours			
Voltage of distribution system	1	10	100	1000
	Percentage of faults restored by time quoted			
33 000 Underground cables	2	6	52	95
Overhead lines	43	83	100	100
Others	20–60	60–80	85–95	100
All	40	72	90	100
11 000 Underground Cables	5	23	90	100
Overhead Lines	20	90	100	100
Others	10–50	65–90	100	100
All	20	80	100	100

(d) Utility share of future national retail market, in detail for five years, in outline for next ten.

In isolated distribution systems only one utility is usually involved. Distribution development then follows selecting the best source(s) and following this through into the future for both new sources and new distribution consumers. A large amount of distribution is unlikely until isolated systems are interconnected either to one another or to the grid. For such isolated systems, load forecasting and market surveys are merged.

7.4.4 Optimum development

Unlike generation and transmission, distribution equipments have shorter lead-times and lives, and thus distribution planning can be done with greater flexibility, and shorter planning horizons, i.e. 3 to 10 years, instead of 20 to 50. Planning of distribution emphasises:

(a) Financial rather than economic data

(b) Short to medium time horizons
(c) Money terms for both costing and pricing, i.e. including inflation
(d) Short-term financial annual returns on assets, not internal returns over lifetimes of equipment
(e) Quick payback periods
(f) Multiple scenario approaches for developments of two or three years.

Table 7.3 illustrates with respect to small systems relative costs of supply for grid extension and isolated generation respectively.

Two complementary approaches are used in planning of optimum distribution developments, both relatively short-term: first, a financial analysis concerned purely with the utility's actual and expected financial performance; and second, a broad economic analysis of overall costs and benefits. The first deals with actual and expected utility money expenditures and returns, examining how costs incurred in implementing alternative distribution development programmes compare with revenues received from them. Projections of expenditures on capital, land, labour, materials, operation and maintenance, interest charges and other expenses to the utility are made for each year to the planning horizon. For programmes which are grid supplied it is assumed that electricity is purchased from generator(s) at the wholesale or bulk supply price. Otherwise something like Table 7.3 must be constructed. Projections are then made of anticipated benefits, mainly of revenues, annually up to the planning horizon

Table 7.3 *Capital cost of public supplies (Source: The author, from World Bank publications)*

Capacity of Scheme	kilowatts	50	50
Distance from grid	kilometre	4	29
Generation and transmission costs	Money units	24 000	24 000
Subtransmission costs	" "	18 000	118 000
Local distribution	" "	14 000	14 000
Total		56 000	156 000
Annual capital costs	" "	5 600	15 600

Load factor %	Supplies from grid		
	4 kilometres	29 kilometres	Autogeneration
10	18	40	21
25	7	17	12
50	4	8	9

Note: Since average costs in urban areas are about 3 cents per kilowatt-hour it is obviously extravagant to extend networks to meet small demands in areas remote from the grid. However, the same subtransmission networks can be used to meet much larger demands. If adequate demand develops from farms, agro-industries, and several villages, average costs decline very quickly to about 4 cents to 8 cents per kilowatt-hour.

and based upon demand forecasts. Using two or more supply options acknowledges different flows of costs and benefits, making it possible to determine when each supply optimum reaches a break-even when revenues exceed costs, up to which point that particular distribution programme would run at a loss, requiring an operating subsidy if chosen. Some programmes never break-even in developing countries. Break-even, however, need not necessarily be defined as meaning profitable to the supplier, because losses incurred before break-even may be unnaturally high, and profits obtained after break-even may be unnaturally low, and it is the total flow of costs and benefits, up to the planning horizon, that must be compared. Some distribution utilities discount these flows to allow for the time value of money, and most allow for inflation.

The above is the financial analysis. The economic analysis compares alternative programmes from the viewpoint of the national economy, to which a distribution utility will pay attention with an interest which depends upon ownership, funding, regulatory obligations, etc. As with the financial analysis, the starting point of the economic analysis is the annual cash flow for each alternative distribution programme being compared, but this time including not simply the utility cash but also all national resources used. For economic analysis, annual financial cash flows are corrected in a variety of ways to represent their true value to society more accurately, e.g. by removing cash subsidies and taxes, putting a resource cost on fuels, labour, capital, etc., in short supply. Economic benefits additional to revenue come from any non-electricity benefits of distribution to agricultural, commercial, or business activity, health, education and welfare. There are difficulties in putting such benefits into cash flow terms, but experience is gradually indicating that these are not as great as accountants often claim. Many distribution programmes with poor financial performance may have an internal economic return which is quite acceptable, given a sufficiently long planning horizon and providing there is a firm programme of government subsidies to the distribution utility in the earlier years.

7.4.5 *Commentary summary*

Distribution systems are normally fed from the grid and used mostly for retail electricity supply, although in the future more private isolated electricity systems may well exist, with diesel, gas turbine or small hydro generators. The most common primary distribution voltage is 33 kV, with secondary voltages 11 kV or 6 kV. Final consumer connections are usually 220/400 volts or 110/200 volts. A variety of possible approaches exist in planning distribution, especially from zero supplies when, with an even distribution of possible consumers, the distribution network will also be evenly spread, ensuring access to supply is uniform and available to all. Another approach is a backbone supply, i.e. a basic distribution line running throughout an area and connections taken from it as demand arises, reducing the level of the initial investment, and the distribution develops in accordance with increasing demands, not earlier. If consumers are 'lumpy' either in size or location, then alternative distribution schemes to provide them must be compared in both financial, utility cash-flow terms, and also national economic terms up to the planning horizon. Any single extensions scheme with respect to existing main distribution must be similarly compared with alternatives. Technically, the choice of single or 3-phase distribution is sometimes important, three phase being more expensive but bringing operating savings and providing greater flexibility and better service to commercial and industrial consumers. Demand forecasting takes on a local intensity for distribution and must include kVar as well as kW and kWh. The level of income in an area is probably the single most important factor determining local electricity demand and the likely success of any distribution development, e.g. in developing countries there is a necessity to determine an income threshold below which people cannot pay electricity connection fees, purchase appliances, or pay electricity usage bills. As disposable incomes rise, consumers have an increasing amount to spend on electricity. For distribution programmes, land and house tenure is also important. People who rent or hold their house or land under temporary or common property rules are less likely to invest in electricity than owner occupiers. Landlords may be willing to install or enlarge electricity supply to increase the rented value. The level of potential demand is influenced by a variety of factors, many difficult to specify let alone quantify, requiring an effective dialogue between all concerned. 'Green field' distribution schemes must compare supply from isolated diesel and/or gas turbine and/or hydro generators with grid extensions. Alternative distribution extension schemes need careful scrutiny to ensure that each is directly comparable. They must then be compared both financially and economically. Annual financial costs to the utility for each alternative scheme are compared with attributable revenues and the best scheme determined. The exercise must then be repeated with the costs and benefits adjusted to reflect shortages and excesses as seen by the national economy, e.g. for shortages of capital, labour etc.

7.5 Utilisation programmes

7.5.1 *Introduction*

Choosing an optimum programme of investment in internal electricity supplies and apparatus starts, as does generation, transmission and distribution planning, with a review of alternative internal supplies and apparatus available, cost, efficiency, reliability, life, etc. The procedure for choosing the optimum development programme is then broadly similar to generation, transmission and distribution, with some industrial nuances [15]. Electricity demand management encourages consumers to adopt cost control options that minimise the need for costly new apparatus for both generator and consumer. Within the comparison of alternative electricity utilisation programmes, reliability standards again play a crucial role.

7.5.2 *Reliability*

Consumers are at risk through:

(a) Loss of electricity supply
(b) Inadequacy of standby supply facilities
(c) Failure of utilisation plant
(d) Inadequacy of standby utilisation plant
(e) Inadequacy of supply quality

Several different reliability levels are normally on offer to consumers, depending on the country, type of system, and the class and type of consumer concerned. Generally the lower the risk the higher the supply cost and the cost of the utilisation apparatus. Especially for medium and large consumers, the option to install standby and cogenerators becomes more attractive the more efficient and competitive the electricity marketplace becomes, especially with dynamic pricing for industrial and commercial consumers. To be effective, such generators must be properly maintained and run periodically over every range of output up to full load. Provision must always be made for adequate fuel storage, fuel store maintenance, and reliable fuel delivery. All these measures add significantly to the cost of standby plant or cogeneration. Standby plant is often similar to normal small generators, but it can be special utilisation plant using other fuels, e.g. gas-oil, coal, petrol. Especially with industrial and commercial consumers with multi-fuel supply policies, utilisation standby plant may use fuels which have implications to the consumer for choice of uses, e.g. heating, motive power and lighting. Whereas there are few substitutes for electric lighting, this is certainly not the case for electric heating and motive power. In both cases, using other energy sources for standby electricity for these tasks can reveal advantages of such energy sources, thus increasing overall the competitiveness of the electricity marketplace. Many industrial and commercial consumers maintain a balance of supply from both the utility and their own cogeneration or autogeneration. Such consumers will be attracted to this in the 1990s because of the increasing efficiency of the electricity marketplaces and it will start applying to large domestic consumers, e.g. blocks of flats, housing estates.

7.5.3 Existing utilisation plant

Existing electricity utilisation plant of any particular firm will depend heavily on existing and previous history of national electricity utilisation plant, international technology trends in utilisation plant, and national technology trends in utilisation plant. The three main uses of electricity are lighting, heating, and motive power. Electrical lighting and motive power efficiency have steadily improved over many years, both holding a unique market which does not change much. However, heating by electricity has undergone considerable technological changes, varying from an outlawing by government and utilities of direct radiant electric fires in the mid 1960s, when every utility was drastically peak lopping, to the overselling of off-peak heating and all-electric factories, commercial establishments and residences in the 1920s and late 1950s/early 1960s. In periods when radiant electric fires are in vogue, tariffs tend to over-charge for electricity consumed at, or around the peak. During the 1930s and late 1950s uneconomic recessionary electricity tariffs, damaging to the sector, were involved whereby the more electricity was consumed, the cheaper it became. Electric heating was regarded somewhat as a misuse of prime energy until the 1980s, the change going on throughout the 1990s due to changes in:

(a) Electricity pricing, away from prescribed towards time-of-use, dynamic and 'spot' pricing
(b) Technology, e.g. in furnaces, the methods of heating, better metallurgy

(c) Understanding of multi-purpose projects, e.g. providing both heat and power, more adequate measuring of attributable costs and separable benefits

(d) Emphasis towards the short-term and financial, flexible type of plant, away from the long-term, economic definitive development programmes as in the past.

Combined heat and power (CHP) plant is covered from the supply side in Chapter 4. Industry has used CHP for some time [16] as part of improving energy efficiency; technology steadily improves and intensive energy users increasingly turn to CHP to reduce fuel bills.

7.5.4 *Expansion of the firm*

The amount and type of any electricity utilisation expansion is affected directly by existing apparatus, amount and type of the firm's products, also whether the latter are likely to change in the short, medium and long-term. Any drastic technological change concerning products is likely to be the most important. Competition for the firm's products is also a vital element in electricity utilisation apparatus development programmes. Coupled with technological changes and expansion of the firm are the opportunities, financial viability of, appropriateness for, and economic worth of substitution into or out of electricity. Multinational firms have an added degree of flexibility because they, besides switching products and fuels, can do these in one country rather than another. It may be technologically appropriate, economic, and financially viable to take advantage of cheap electricity in, say, a predominantly hydro system in making aluminium, rather than in a predominantly thermal system with inevitably dearer electricity.

7.5.5 *Pricing*

Pricing schedules in which tariffs change periodically and possibly in real-time are the subject of increasing discussion from the demand as well as the supply side, different tariffs per kWh being requested during different, pre-specified periods of the day, week, and year (Chapter 6). Such time-of-use tariffs have been implemented in various parts of the world. From the consumers' point of view, time-of-use tariffs are superior to uniform tariffs because consumers are more likely to pay for the marginal costs of the electricity they use. However, prescribing prices well in advance can never achieve a consumer or national welfare optimum, since actual demand at any future time will vary considerably from average. Dynamic pricing (Chapter 6) addresses this from the national and the utility's viewpoints. An important issue from the demand side is consumer response to these rapidly changing prices. However, those for whom there are most risks could buy electricity forward or conclude supply options for the future with generators, the grid, distribution utilities using wheeling and agents or brokers. By charging a premium for buying forward, or for options for future supplies (i.e. over the expected spot electricity price at the time of electricity usage) utilities could make sure of some consumers' electricity usage.

7.5.6 Commentary summary

The traditional methodologies for determining optimum utilisation development programmes are broadly similar to those for determining optimum generation, transmission and distribution programmes, but the choice of economic electricity supplies must be extended more towards using internal small generators, cogenerators, autogenerators and combined heat and power. Optimising utilisation plant programmes must always be basically tied to the consumer's own activities and products. However, demand management and dynamic or spot pricing must directly affect these, altering the balance of equipment to suit both electricity suppliers and consumers.

7.6 References and further reading

1. CORY, B.J. (1986) 'Technology 2000', paper for *Technology 2000 Exhibition* (Imperial College, London); also, Cory, B.J. (1988) 'Generating plant', paper to *Planning of Small Systems Course* (Imperial College, London)
2. WEBB, M. (1979) 'Electricity sector planning manual' (Overseas Development Administration, London); also Bridger, B.A. and Winpenny, J.T. (1983), 'Planning development projects' (Overseas Development Administration, London) Chapter 8 especially; also, Berrie, T.W. (1983) 'Power system economics' (Peter Peregrinus)
3. OWEN, B. (1956) 'Thermodynamics' (Osman Press, USA) or any textbook on thermodynamics
4. ONGMACH, P. (1987) 'The status of geothermal resources plant', *IEE Conference on the Future of Fuels*, London
5. FRERARIS, B. (1988) 'Renewables': paper for *Course on Small Electricity Systems* (Imperial College, London)
6. CARUTHERS, R. (1988) (discussion leader) 'Session 12 Fusion': paper for *Course on Small Electricity Systems* (Imperial College, London)
7. BERRIE, T.W. (1967) 'Public enterprise,' Chapter 5; (Penguin Economics); also, Berrie, T.W. (1981) 'Power system planning under uncertainty', *Power Engineering Journal*, **1**, (2), pp. 92–100; also, Berrie, T.W. (1983) 'Power system economics' (Peter Peregrinus) Chapter 3; also, Munasinghe, M. (1979) 'The economics of power system reliability and planning' (The Johns Hopkins University Press) Chapter 3 and 4; also, Munasinghe, M. (1990) 'Electric power economics' (Butterworths); also, Grubb, V.J. (1985) Probabilistic electricity generation analysis', Energy and Power System Report, No. 112 (Imperial College, London)
8. BERRIE, T.W. (1988), 'The new power sector planning', *Energy Policy*, October, pp. 453–457
9. BERRIE, T.W. and ANARI, M.A.M. (1986) 'Joint supply-demand optimisation in electricity supply', *Energy Policy*, December, pp. 515–527
10. BERRIE, T.W. (1983), 'Power system economics' (Peter Peregrinus Ltd., London) Chapter 10, this also giving many references
11. CORY, B.J. (1988) 'Transmission planning', for *Course on Planning Small Systems* (Imperial College, London)
12. CORY, B.J. (1988) 'Use of system for wheeling', paper for *Course on Planning Small Systems* (Imperial College, London); also, Schweppe, F. *et al.* (1988) 'Electricity spot pricing' (M.I.T. Kluwer Press)
13. BERRIE, T.W. (1983), 'Power system economics', Chapter 10, section 10.5; also, Munasinghe, M. (1990) 'Electric power economics', (Butterworth Heinemann), pp. 284–314
14. FOLEY, G. (1990), 'Electricity for rural people', Panos Rural Electrification Programme 1990, Panos Institute, 8 Alfred Place, London; also, Munasinghe, M. (1979) 'The economics of power system reliability and planning' (The Johns Hopkins University Press); also, Munasinghe, M. (1987), 'Rural electrification for development policy analysis and applications' (Westview Press, Boulder, Colorado, USA)

15 HOSE, T.F. (1986) 'Demand-side planning: a practical perspective', paper to *Winter Meeting IEEE Power Engineering Society*, New York
16 WOOD, B. (1970) 'Alternative fluids for power generation', *Proceedings Mechanical Engineers*, **194**, Part 1 (40) pp. 713–740; also, Hellemans, I.R.J., (1978) 'European practice in self-generation of electricity conference: local energy centres', ed. N.J.D. Lucas (Applied Science Publishers, London) pp. 44–61

7A APPENDIX
TECHNOLOGY TRANSFER

7A.1 Introduction

Technology transfer, crucial to electricity supply on both supply and demand sides, is advancing increasingly rapidly, especially through electronics and telecommunications, new materials, solid state physics and scientific instruments. Particularly, computer-driven industrial and control systems, plus faster technology diffusion, have become universal. Worldwide, automatised technologies requiring much electrical energy are rapidly replacing labour, needing specialist skills, management and techniques, e.g. 'just-in-time' procedures. Developing countries find technology advancing so rapidly that they must import rather than produce internally. To obtain technological capability, incentives are needed to develop institutions and information resources. Sometimes it is best overall to continue with traditional technologies temporarily, sometimes to adopt the most modern immediately. Once transferred, technology must be assimilated and diffused properly by geography, scale and situation, involving training, after which local technologies can often be used alongside imported ones. All transfers to be effective must be legal, nationally and internationally, initially involving royalty payments. Practical economies of scale must be easily discernible because wrong choices bring high penalties. Technology transfers to be effective require technical assistance given by and to open but discerning minds which are both motivated for change.

7A.2 Transfer process

Important technology transfer theories of the 1970s lost influence with falling oil prices. More important today is answering the perennial questions: what is appropriate, what are transfer benefits and who gains? Funds and expatriate expertise, transfer benefits, short and long-term, must also be considered. Technologies, resources, skills, etc., developed for one country cannot be used automatically in another, taking guidance from the relative per capita income of donor and receiver, possibly suggesting a staged transfer. All countries possess latent skills easily identifiable in developed but not in developing countries. These skills must be gradually brought out into eventual full partnership with the technology transfers. Transfer success means having the right donor and recipient institutions, possibly requiring special institutions set up to cut through bureaucracies, encourage joint private-public investorships and allocate scarce resources. Transference places most risks with the recipient even though donors are better placed for risk. Apportioning risks under high financial stakes requires great skills, such as those of the World Bank.

7A.3 Appropriate technology

Mostly complex, sophisticated technology is transferred but there is growing belief in spreading the investment available for electricity across much broader cross-sections of transfer types. In the 1960s, simpler technology transfers were popularly advocated with little effect, except for marginal trimming of design standards of advanced technologies to reduce costs. Factors when seeking appropriateness are: national objectives and economic plans; past experience; return on resources; institutional gains; and fitting technologies into existing environment. Compromises must be known and reasonable. Rural electrification is the most important candidate for appropriate technology and research is still needed. There may be local manufacturers, especially in urban areas, to help fit in appropriate technology because appropriateness may mean at first using labour rather than capital-intensive technologies, although this is unlikely to be economically successful long-term, long-term appropriateness depending more on marketplace efficiency in labour, services, finance, and goods.

Research is still needed within the electricity sector examining issues and options involved in technology transfer in particular circumstances. Large savings may indeed be forthcoming because of the sector's extreme capital intensity. Renewables must be keyed into this research. In the developing countries, manufacturers, consumers, and even workers are in a type of low-grade symbiosis, looking for an escape.

7A.4 Innovation

Successful innovation requires a fertile imagination acting within an innovative environment, wide enough to include national economic energy and electricity policies, objectives, markets and pricing. The favoured innovative process is top-downwards, trickle down, large to small, skilled to unskilled. To transfer technology successfully requires a critical existing level of technical, financial, social and institutional resources. Successful innovation quickly spots the level needed and is adaptable to the economy, culture, sector, utility and consumers, fully using all innovative advantages. Innovation works best when forming an intrinsic part of planning because this indicates credible scale, scope and timing. Innovation must seem to fit well into the particular economy, sector and utility market, as seen by receivers. Success requires adequate external and internal backing, sufficiently not involved in transfer day-to-day problems.

Within the background into which it is introduced, innovation must be clearly prescribed and understood, what it does, with what consequences, sectors affected etc. Seeing is believing; demonstrating the innovation will work is as important as explaining what it does, but good innovators are seldom good explainers, and greatest emphasis is needed at the early stages, when expatriate roles can be paramount. Some innovations prove effective only at particular sizes, especially in initial stages (e.g. steam generators, supergrids), or funding package size (e.g. for distribution). Again, successful innovation means appropriate, adequate training, an integral part of the innovative process. In this sense training is a specialist function.

7A.5 Scale economies

The years following World War II were challenging to technology transfer in electricity supply, with design frontiers short-lived and both size and operating conditions undergoing rapid change under the stimuli for scale economies (Figures 7.4, 7.5 and 7.6 refer). Starting in developed countries, this progression has long since spread to developing countries. However, the promised capital savings and high efficiencies often were not fully realised. In the UK, the first 200 MW generator was ordered before the first 120 MW unit operated. Again, the first 500 MW generators were ordered before the first 275 MW units operated. This also happened in other countries, placing general strain on designers, manufacturers, contractors, and operators. It seems doubtful in the 1990s whether further pursuance of such scale economies in developing countries is worthwhile. Nor does there seem room for developed countries pursuing further scale economies in generation or transmission, especially

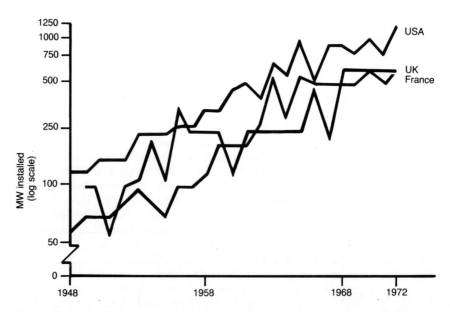

Figure 7.4 *Size of largest turboalternator installed in the year (Source: TOMBS, F. 'Economies of scale in electricity generation and transmission since 1945', Proceedings of the Institution of Mechanical Engineers, Vol. 192, 1978, Figure 3. Reproduced by permission of the Chartered Mechanical Engineer)*

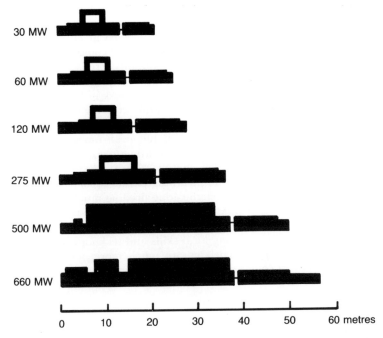

Figure 7.5 *Size of generating sets (Source: TOMBS, F. 'Economies of scale in electricity generation and transmission since 1945', Proceedings of the Institution of Mechanical Engineers, Vol. 192, 1978, Figure 5. Reproduced by permission of the Mechanical Engineer)*

bearing in mind the change in emphasis in favour of smaller plants. The task of delivering vast amounts of electricity from the increasingly large generators in the past fell to the transmission grids. Again economies of scale were sought, through higher voltages, larger circuits, transformers, switchgear, etc. Distribution followed transmission. Larger electricity projects meant larger civil works, taking many years to construct. Important economies of scale during construction can be achieved by the intensive use of machinery to replace labour. However, this is challengeable in socio-economic terms in developing countries, and more research is needed to quantify all the social costs within the comparison.

7A.6 Training

The volume of skills transferred per unit of time and effort is proportional to the difference in knowhow between trainer and trainee, and inversely proportional to the resistance encountered by the latter's background, standard of living, psychology and language. Within all these factors, finding the right training method is paramount, resistance to technology transfer affects its pace but resistance is lowered by training of everyone involved at *all* levels. Training agencies are: government departments; manufacturers; contractors and utilities who are part of the transfer process; consultants preparing the transfer project(s); and any local government departments, agencies, consultants, manufacturers and contractors involved. *All* must work in harmony.

For any technology transfer to be successful, the receiving country and utility must provide adequate teams to work alongside donors. Local people involved must be clearly designated, reasonably available at all times to give their full attention to the transfer, with proper communication to donors. With respect to training for this end, recommended methods vary considerably from the entirely formal to complete on-the-job learning packages, depending upon the discipline and the project involved, engineering needing much more on-the-job training than economics and finance. Perhaps technologists have the worst reputation worldwide for using training methods and objectives above the required level of recipients, ignoring local skills and technologies, ignoring the fact that mistakes made by expatriates are much more noticeable and less correctable than those of compatriots. Ability to make independent technologi-

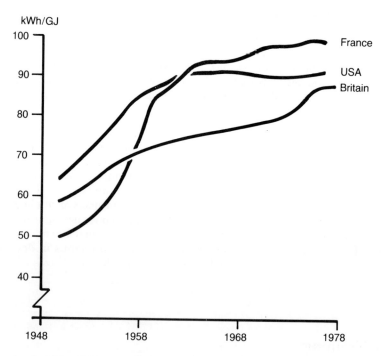

Figure 7.6 *Fuel utilisation (Source: TOMBS, F. 'Economies of scale in electricity generation and transmission since 1945', Proceedings of the Institution of Mechanical Engineers, Vol. 192, 1978, Figure 4. Reproduced by permission of the Chartered Mechanical Engineer)*

cal choices and improve upon chosen techniques and operation, eventually to generate new technology endogenously, are essential aspects of technology transfer, at least as important to economic development as the accumulation of capital and other resources.

7A.7 Patents

Most industries, infrastructures and all electricity sectors heavily depend upon patent rights, about which there is much ignorance. Patents costs can add considerably to transfer costs, possibly preventing transfer. International patents were first considered globally in the 1960s; each geographical group of nations has its own patent rights, but needed are explicit international patent rights, registering of patents, etc., e.g. through the United Nations Conference on Trade and Development.

The two aspects still needing addressing are patent format and administration. Patent formats are slowly becoming more comparable worldwide, despite language difficulties, spelling out required degrees of detail, patent access, administration, etc. Patent administration must include disputes procedures, claims handling, patent enforcement, compensation and guarantees. Each nation has patent laws but there are three interwoven threads of generally applicable legal obligations: enforcement through national and international courts; government action, especially with respect to international conglomerates, suppliers, manufacturers, etc.; and national regulatory control of usage and abusage of patents, local and foreign. Besides the UN there are rapidly growing numbers of international agencies funding involvement with patents, e.g. the World Bank, EEC, World Energy Conference, national and international environmental agencies.

7A.8 The future

In the 1990s economic growth must rely on sound technology and there already exists a distinct progression worldwide from lower to higher technology. Technology transfer is, therefore, likely to speed up in sectors which heavily depend on it, e.g. electricity. Many types of technology transfer take place daily to electricity sectors, some still proving unsuccessful. Given the changes taking place nationally and internationally in electricity, there should be a gradual improvement in its transfer process. Electricity can learn perhaps from the rapid development of the electronics sector. In 1950 this was virtually non-existent except in the USA. In 1957 world electronics production barely exceeded US$ 121 billion, still mostly in the USA. Between 1957 and 1972 the growth rate of world electronics production was steady at about 13% per year. However, it started speeding up about 1974 and averaged 16% per year between 1974 and 1982, remarkable considering the fall in costs in information processing and the world economy slowed dramatically after 1974. A 1985 memory 'bit' cost 0·5% of what it had in 1970. The key was in technology development and transfer.

7A.9 Further reading

A1 DAHLMAN, C.J. (1987) 'Technology change in industry in developing countries'. *Finance and Development*, June, pp. 13–15; also, Berrie, T.W., (1980) Series on technology transfer in United Nations Development Forum Business Edition; also, Schumacher, E. (1971), 'Small is beautiful' (Oxford University Press)

A2 STEINBERG, R.M. (1977), 'Devising a system for appropriate technology', World Bank Report, July–August; also Rotberg, R. (1987), 'Why not try some other technologies' (Goldberg PLC Publishers, Holland)

A3 Research reports of the World Bank on technology transfer

Chapter 8
Developing countries

8.1 Introduction

8.1.1 *Special features*

Special features of particularly the poorest developing countries are very apparent. Some have already been mentioned [1], e.g. population growth (Chapter 1 and Figure 8.1) is a basic influence on per capita income and, therefore, electricity demand growth. The rapid developing countries' population growth, which is unlikely to slacken off in the near future, is in contrast to developed countries reaching population saturation. Developed countries' experiences indicate that there is a low-income trap which only huge capital formation, or major technology transfer stimuli, or both, can avert. Population changes being both consequences and causes of economic change, their effects are hard to trace and quantify. Partly because of this, many dealing with electricity demand and consumer classes in developing countries prefer to work with disaggregated population data (see Table 8.1) available from regional UN agencies. Others deal with average growth in total population by sub-continent, at least as a starting point for cross-sectional analysis and extrapolation (Figures 8.2 and 8.3).

8.1.2 *Economic growth*

Likely per capita income growth is another factor in forecasting developing country electricity demand. Development growth is as complex as is population growth, with as few short cuts in estimating it. However, there are some general lessons to be learned about growth from literature scans, e.g. sustained economic growth needs high investment and import growth, also the ratio of developing to developed countries' exports in 1987 was almost the same as in 1967. Differing regional patterns of growth can be traced to nature, history, politics and level of debt, but making cross-sectional analysis and forward projections is well worthwhile (Tables 8.2, 8.3, 8.4 and 8.5).

Each developing country needs to amass information and carry out analyses particular to that country. Indications overall for the region in which the developing country falls are obtainable from well published material from the World Bank and its sister Banks, e.g. Inter-American and Asian Development Bank. From these tables some indications are noteworthy: the gap between the world's rich and poor has widened during the past two decades; overall income growth in low and middle income developing countries peaked in 1982 with per capita earnings of about US$760 per year (see Tables 8.6 and 8.7). Some world regions were worse off in 1990 than two decades before, reference the GDP per

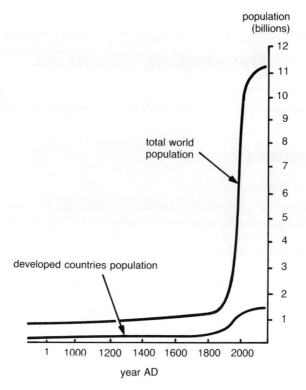

Figure 8.1 *Past and projected world population AD1–2150 (Source:* Research News, the World Bank, Vol. 9, No. 1, September 1989)

capita in sub-Saharan Africa (except Nigeria) dropping from US$400 in 1968 to US$340 in 1988. Also, during the past twenty years, private consumption per capita remained unchanged in almost all developing countries. This is important for electricity forecasting in that private consumption measures a person's ability to purchase basics, food, shelter rent, health, education, consumer and other goods and services, including electricity. Those countries having gains in per capita private consumption tended to record gains also in per capita investment. Gross domestic investment per capita for most developing countries (Table 8.8) peaked in the early 1980s, then fell, echoing the pattern of GDP per capita. Because electricity is capital-intensive, it is vital to forecast likely future investment available, local and foreign.

Electricity is sensitive to income, especially for domestic and commercial consumers. Most past attempts by governments to redistribute income have had little success, e.g. more than 1 billion people still live in poverty, with individual incomes below US$ 370 per year [2]. However, the number of poor people will decrease if developing countries pursue policies of encouraging the use of their large amounts of labour and providing adequate social services,

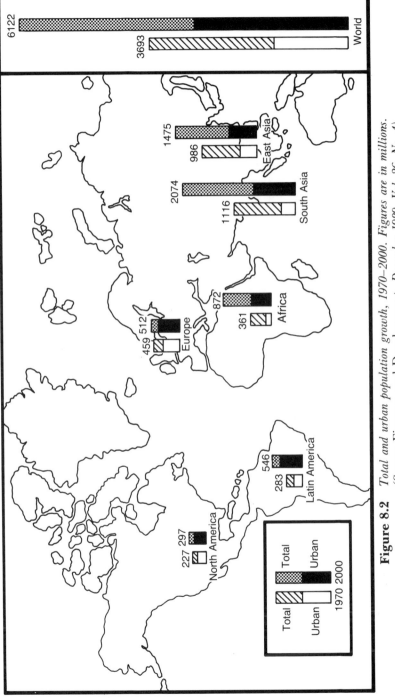

Figure 8.2 *Total and urban population growth, 1970–2000. Figures are in millions.* (*Source:* Finance and Development, December 1989, Vol. 26, No. 4)

Table 8.1 Some social indicators of development in Africa (Source: Various World Bank published economic reports entitled 'Economic conditions and current prospects', World Bank, Washington DC USA)

	Crude birth rate	Crude death rate	Infant mortality rate (live birth)	Life expectancy at birth (Years)	Gross reproduction rate	1975 Population				Age Structure			Dependency ratio
						Rural	Urban	% Urban		0–14 Years	15–64 Years	65 Over	
	per 1000					million				%			
Chad	43	23	160	40	—	3·46	0·56	13·9		41	55	4	0·8
Gambia	43	23	65	41	2·8	0·43	0·08	15·0		41	55	4	1·0
Mali	50	27	—	37	3·3	4·93	0·77	13·5		49	49	2	1·0
Mauritania	44	23	169	41	2·9	1·14	0·14	11·0		44	51	5	1·0
Niger	52	22	162	43	—	4·16	0·43	9·4		45	51	4	0·9
Senegal	46	23	156	—	3·0	3·17	1·25	28·2		42	54	4	0·9
Upper Volta	49	29	182	35	3·2	5·53	0·50	8·3		43	54	3	0·9
Sahelian						22·82	3·73	14·0					
Guinea	47	25	216	—	—	3·56	0·86	19·5		44	47	9	1·1
Guinea Bissau	41	30	47	34	—	0·40	0·12	23·1		—	—	—	—
Liberia	50	21	159	53	2·6	1·30	0·41	24·0		42	55	3	1·2
Sierra Leone	45	23	183	41	2·9	2·44	0·54	18·0		42	55	3	1·1
Coastal West						7·70	1·93	20·0					
Benin	54	26	110	—	—	2·52	0·15	17·9		46	48	6	1·1
Ghana	47	18	155	47	3·2	6·41	3·26	35·0		47	49	4	1·3
Ivory Coast	46	23	140	42	3·1	4·24	2·10	34·1		42	55	3	1·0
Nigeria	49	25	157	39	3·3	59·00	16·60	22·0		45	53	2	1·2
Togo	31	29	179	40	—	1·84	0·40	18·0		47	47	6	0·9

Developing countries 217

| | | | | | | 74.01 | 23.21 | 23.9 | | | |
						104.53	28.87	27.62			
Coastal East Western Africa											
Cameroon	—	—	142	48	—	4.88	1.52	23.9	56	5	0.8
CAE	—	—	175	40	—	1.11	0.67	37.6	54	4	0.9
Congo	44	21	160	42	—	0.81	0.53	39.6	46	5	1.0
Eq. Guinea	35	22	53	41	—	0.24	0.08	25.0	82	4	—
Gabon	33	25	—	39	2.0	0.36	0.13	26.5	61	5	0.6
Equatorial						7.40	2.93	28.4			
Angola	—	—	203	35	—	4.92	0.55	10.0	54	4	0.9
Burundi	41	20	138	41	2.8	3.66	0.07	2.0	52	3	0.9
Rwanda	50	22	163	43	—	4.24	0.12	3.0	46	3	1.2
Zaire	44	23	104	41	2.8	18.05	6.68	27.0	55	3	1.1
Eq. Central Central Africa						30.87	7.42	19.4			
						38.27	10.35	21.28			
Botswana	45	14	126	55	3.2	0.56	0.11	17.0	49	5	1.0
Ethiopia	46	25	84	39	1.0	24.85	3.07	11.0	51	4	1.1
Kenya	50	17	—	50	—	11.61	1.74	13.0	46	5	—
Lesotho	37	15	106	46	1.1	1.17	0.04	4.0	55	4	0.8
Madagascar	—	—	—	—	—	7.69	1.14	13.0	—	—	—
Malawi	54	28	142	39	3.2	4.68	0.41	8.0	52	4	0.9
Mauritius	25	8	164	63	2.3	0.86	0.03	3.0	56	4	0.9
Mozambique	43	23	93	41	2.8	8.13	1.11	12.0	56	4	1.5
Somalia	47	22	—	40	3.0	3.15	0.03	1.0	52	2	0.9
Swaziland	52	24	—	41	3.5	4.45	0.49	10.0	49	3	1.0
Tanzania	47	22	163	43	3.2	13.56	1.17	8.0	53	3	0.9
Uganda	45	17	136	49	—	10.51	1.04	9.0	48	5	1.1
Zambia	50	21	—	44	3.3	3.59	1.33	27.0	51	3	1.3
Eastern Africa						94.81	11.71	11.0			
Africa						237.61	50.93	17.65			

218 *Developing countries*

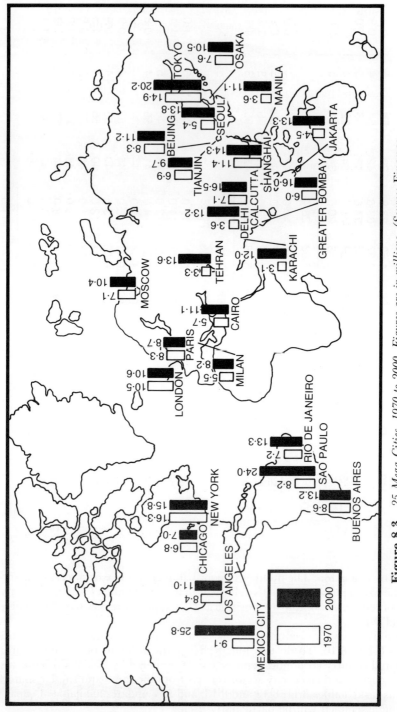

Figure 8.3 *25 Mega Cities, 1970 to 2000. Figures are in millions.* (*Source:* Finance and Development, *December 1989*, Vol 26, No 4)

Table 8.2 *GDP growth rates in low and middle income countries (Annual average percentage change. Source:* World Bank News, *August 9, 1990, p. 2)*

	1989	1990	1991
Low- and middle-income countries	3.4	3.2	4.2
By region			
Sub-Saharan Africa[a]	3.2	2.9	3.3
Nigeria	5.3	3.7	3.7
Other Sub-Saharan Africa	1.7	2.3	2.9
Eastern and Southern	2.7	2.4	3.6
Western	0.3	2.2	1.9
Asia	5.2	4.9	5.6
East Asia	5.4	5.0	5.8
China	3.9	3.5	5.0
Others	7.4	6.9	6.7
South Asia	4.7	4.9	5.1
India	4.8	5.0	5.3
Others	4.1	4.7	4.6
Europe, Middle East and North Africa	1.6	0.8	2.2
Eastern Europe[b]	0.2	−3.4	−0.1
Others	2.3	3.0	3.4
Latin America and Caribbean	1.5	1.6	3.0
Brazil	3.5	0.0	2.0
Other Latin America	0.4	2.5	3.6
By income groups			
Low-income countries	4.4	4.0	4.9
Large low-income countries	4.2	4.0	5.1
Small low-income countries	5.1	4.3	4.4
Middle-income countries	2.7	2.6	3.7

Note: Gross domestic product (GDP) is measured and aggregated in constant dollars at 1980 prices and exchange rates.
[a] Excludes South Africa
[b] Comprises Hungary, Poland, and Yugoslavia

especially education. The number of poor will increase in world recession, and/or if the world pursues restrictive short-term financial policies, e.g. in restrictive trade practices. An optimistic position is shown in Table 8.8, based on Tables 8.9 and 8.10.

There is always a need to assemble figures similar to Figures 8.1 to 8.6 and Tables 8.1 to 8.10. To do this much basic data needs to be assembled from world and country economics reports, e.g. as prepared by the World Bank; also similar energy and electricity sector reports from the same and similar sources. For this task, Tables 8.11, 8.12 and 8.13, and any updates available or capable of being assembled will be particularly useful.

Table 8.3 *Per capita GDP growth rates in low and middle income countries (Annual average percentage change. Source:* World Bank News, *August 9, 1990 p. 3)*

	1989	1990	1991
Low- and middle-income countries	1·4	1·2	2·2
By region			
Sub-Saharan Africa[a]	0·0	−0·3	0·1
Nigeria	2·2	0·7	0·7
Other Sub-Saharan Africa	−1·5	−0·9	−0·3
Eastern and Southern	−0·5	−0·9	0·3
Western	−2·8	−1·0	−1·1
Asia	3·3	3·0	3·8
East Asia	3·7	3·3	4·2
China	2·4	1·9	3·6
Others	5·4	5·0	5·0
South Asia	2·4	2·7	3·0
India	2·7	2·9	3·3
Others	1·4	2·0	1·9
Europe, Middle East and North Africa	−0·3	−1·0	0·4
Eastern Europe[b]	−0·3	−3·8	−0·5
Others	−0·1	0·7	1·0
Latin America and Caribbean	−0·6	−0·4	1·1
Brazil	1·4	−1·9	0·2
Other Latin America	−1·6	0·4	1·6
By income groups			
Low-income countries	2·3	1·9	2·9
Large low-income countries	2·4	2·2	3·4
Small low-income countries	2·3	1·5	1·7
Middle-income countries	0·7	0·6	1·8

Note: Gross domestic product (GDP) is measured and aggregated in constant dollars at 1980 prices and exchange rates.
[a] Excludes South Africa
[b] Comprises Hungary, Poland, and Yugoslavia

Table 8.4 *GDP growth rates in low and middle income countries (Annual average percentage change. Source:* World Bank News, *March 15, 1990 p. 2)*

	1988	1989	1990	1991
Low- and middle-income countries	5·5	3·5	3·6	4·7
Sub-Saharan Africa	2·8	3·2	2·4	3·2
Asia	9·7	5·4	5·4	6·5
Europe, North Africa, Middle East	2·7	1·8	1·0	3·1
Latin America and Caribbean	1·0	1·2	2·3	2·7

Table 8.5 *GDP per capita growth rates in low and middle income countries (Annual average percentage change. Source:* World Bank News, *March 15, 1990, p. 2)*

	1988	1989	1990	1991
Low- and middle-income countries	3.4	1.5	1.6	2.7
Sub-Saharan Africa	−0.5	0.0	−0.7	0.0
Asia	7.7	3.5	3.5	4.7
Europe, North Africa, Middle East	0.8	−0.1	−0.9	1.2
Latin America and Caribbean	−0.1	−0.8	0.2	0.8

Table 8.6 *Gross national product per capita, current US$ (Source:* World Bank News, *April 19, 1990, p. 5)*

Region	1968	1978	1988
Sub-Saharan Africa	120	400	330
South Asia	100	180	320
East Asia and Pacific	100	310	550
Latin America and Caribbean	490	1460	1860
Middle East and North Africa (excluding Iran and Iraq)	220	750	1210

Table 8.7 *Gross domestic investment per capita (1980 US$) (Source:* World Bank News, *April 19, 1990, p. 6)*

Region	1968	1978	1988
Sub-Saharan Africa	60	120	60
South Asia	40	50	60
East Asia and Pacific	40	120	240
Latin America and Caribbean	250	460	330
Middle East and North Africa (excluding Iran and Iraq)	120	360	300

8.1.3 *Externalities*

Factors external to electricity became important to developed countries about ten years ago, e.g. demand management, efficiency, conservation, and environmental maintenance. Some argue [3] that only industrialised countries can afford these factors which, after all, require even more finance to be found by developing countries, already too heavily in debt. Nevertheless, agencies such

Table 8.8 *Poverty in 2000, by developing region (Source:* World Bank News, *July 19, 1990, p. 2)*

Region	Incidence of Poverty		Number of Poor (millions)	
	1985	2000	1985	2000
Sub-Saharan Africa	46·8	43·1	180	265
East Asia	20·4	4·0	280	70
China	20·0	2·9	210	35
South Asia	50·9	26·0	525	365
India	55·0	25·4	420	255
Eastern Europe	7·8	7·9	5	5
Middle East, North Africa and Other Europe	31·0	22·6	60	60
Latin America and the Caribbean	19·1	11·4	75	60
Total	32·7	18·0	1125	825

Note: The incidence of poverty is the share of the population below the poverty line, which is set at $370 annual income

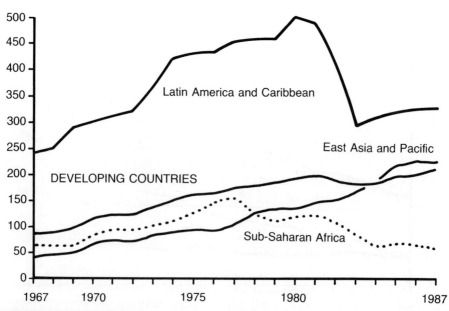

Figure 8.4 *Gross domestic investment per capita 1967–87 in US dollars (Source:* World Bank News, *May 1989, p. 6)*

Table 8.9 *Performance indicators: selected periods (Source:* World Bank News, *July 19, 1990, p. 3)*

Region	Growth of Real Per Capita GDP (percent)		
	1965–73	1973–80	1980–89[a]
Sub-Saharan Africa	3·2	0·1	−2·2
East Asia	5·1	4·7	6·7
South Asia	1·2	1·7	3·2
Eastern Europe[b]	4·8	5·3	0·8
Middle East, North Africa and Other Europe	5·5	2·1	0·8
Latin America and the Caribbean	3·7	2·6	−0·6

[a] Data for 1989 are preliminary
[b] Estimates

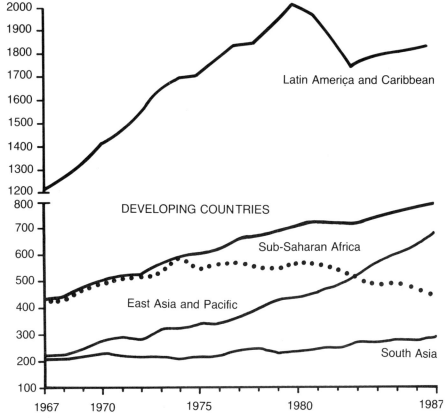

Figure 8.5 *Gross national income per capita 1967–87 in US dollars (Source:* World Bank News, *May 1989, p. 7)*

Table 8.10 *Real GDP per capita growth rates (Source:* World Bank News, *July 19, 1990, p. 4)*

Trend Group and Region	Recent experience 1965–1980	Forecast 1980–1989	1989–2000
Industrial Countries	2·8	2·5	2·6
Developing Countries	3·4	2·3	3·2
Sub-Saharan Africa	2·0	−2·2	0·5
East Asia	4·8	6·7	5·1
China	4·1	8·7	5·4
Other	5·5	4·2	4·6
South Asia	1·2	3·2	3·2
India	1·2	3·5	3·4
Other	1·2	2·2	2·4
Eastern Europe	4·5[a]	0·8[a]	1·5

Middle East, North
  Africa and Other
  Europe
	3·9	0·8	2·1

Latin America and the Caribbean	3·4	−0·6	2·3

[a] estimates

as the World Bank have more recently grown aware of these socio-economic impacts of development and have made lending goals more in harmony with nature. The extra monies forthcoming for this purpose in the past had depended on individual projects and donors, but the World Bank and others are now making special allocations. Electricity sectors are thus expected to give proper attention to these externalities, even in developing countries.

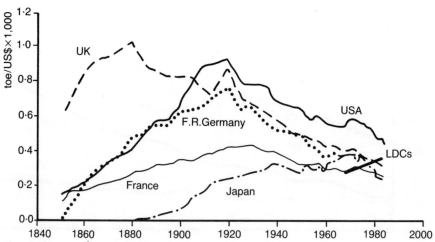

Figure 8.6 *Evolution of the energy intensity in different countries (Source:* Oxford Energy Forum, *Vol. 1, No. 2, August 1990, p. 3. Reproduced by permission of Revue de l'Energie, Paris)*

Table 8.11 Three measures of GDP per capita in 1985 (Source: 'The rich and the poor: changes in incomes of developing countries since 1960', Overseas Development Institute, London, 1988, Table 1. Reproduced by permission of the Overseas Development Institute, London)

	1 International or real national accounts 1980 international dollars		2 At 1980 prices and exchange rates conventional national accounts		3 At 1985 prices and exchange rates
			US$		US$
Norway	12623	Norway	16413	US	16759
US	12532	Germany, FR	14235	Norway	13799
Canada	12196	France	12973	Canada	13431
Germany, FR	10708	US	12843	Japan	11014
France	9918	Belgium	12563	Australia	10481
Belgium	9717	Netherlands	12285	Germany, FR	10270
Japan	9447	Canada	11633	France	9466
Hong Kong	9093	Austria	11004	Austria	8692
Netherlands	9092	Australia	10968	Netherlands	8692
Austria	8929	Japan	10636	Belgium	8186
Australia	8850	UK	10384	UK	8127
UK	8655	Italy	8647	Italy	6224
Italy	7425	Hong Kong	6472	Hong Kong	6170
Spain	6437	Ireland	5947	Israel	5404
Israel	6270	Spain	5937	Ireland	4600
USSR	6266	Israel	5863	Spain	4204
Hungary	5765	Argentina	4599	Greece	3339
Ireland	5205	Greece	4315	Taiwan	3097
Yugoslavia	5063	Venezuela	3204	Venezuela	2750
Poland	4913	USSR	3037	USSR	2449
Greece	4464	Taiwan	2871	Argentina	2153
Mexico	3985	Yugoslavia	2818	South Korea	2039
South Africa	3885	South Africa	2591	Hungary	1936

Table 8.11

1 International or real national accounts 1980 international dollars		2 At 1980 prices and exchange rates conventional national accounts		3 At 1985 prices and exchange rates	
Portugal	3729	Portugal	2576	Yugoslavia	1913
Taiwan	3581	Mexico	2569	Portugal	1909
Venezuela	3548	Hungary	2270	Poland	1894
Argentina	3486	Chile	2231	Mexico	1872
Chile	3486	South Korea	2192	Malaysia	1844
Malaysia	3415	Brazil	2028	South Africa	1600
Brazil	3282	Malaysia	2005	Brazil	1585
South Korea	3056	Tunisia	1404	Ecuador	1578
Colombia	2599	Ecuador	1392	Guatemala	1363
Turkey	2533	Turkey	1388	Chile	1320
Ecuador	2387	Colombia	1295	Colombia	1160
Peru	2114	Ghana	1126	Tunisia	1151
Tunisia	2050	Cote d'Ivoire	1041	Turkey	1069
Thailand	1900	Guatemala	935	Bolivia	795
Guatemala	1608	Morocco	932	Nigeria	768
Sri Lanka	1539	Peru	847	Thailand	718
Philippines	1361	Thailand	832	Peru	708
Indonesia	1255	Zimbabwe	820	Cameroon	698
Morocco	1221	Nigeria	793	Cote d'Ivoire	641
Egypt	1188	Cameroon	791	Philippines	584
Pakistan	1153	Bolivia	691	Zimbabwe	583
Cameroon	1095	Egypt	621	Morocco	552
Bolivia	1089	Philippines	611	Egypt	529
Zimbabwe	948	Zambia	607	Indonesia	499
Cote d'Ivoire	920	Indonesia	552	Ghana	461
China	825	Kenya	399	Zambia	389

Senegal	754	Pakistan	339	Haiti	377
India	750	Madagascar	318	Sri Lanka	376
Kenya	698	Sri Lanka	313	Pakistan	337
Bangladesh	647	India	282	Tanzania	283
Haiti	631	China	273	Kenya	277
Zambia	584	Haiti	259	India	261
Nigeria	581	Tanzania	247	Madagascar	235
Madagascar	495	Zaire	215	China	222
Malawi	387	Malawi	193	Malawi	159
Mali	355	Bangladesh	157	Bangladesh	152
Tanzania	355	Ethiopia	112	Ethiopia	111
Ghana	349			Zaire	79
Ethiopia	310				
Zaire	210				

228 *Developing countries*

Table 8.12 *Changes in national income over time (Source: 'The rich and the poor: changes in incomes of developing countries since 1960', Overseas Development Institute, London, 1988, Table 2. Reproduced by permission of the Overseas Development Institute, London)*

	1960	1965	1970	1975	1980	1985
Income per capita (1980 international dollars) Above 7,400	US	US	Canada Germany, FR US	Australia Belgium Canada France Germany, FR Netherlands Norway US	Australia Austria Belgium Canada France Germany, FR Japan Netherlands Norway UK US	Australia Austria Belgium Canada France Germany, FR Hong Kong Italy Japan Netherlands Norway UK US
2,300–7,400	Argentina Australia Austria Belgium Canada Chile France Germany, FR Hungary Ireland Israel Italy Netherlands	Argentina Australia Austria Belgium Canada Chile France Germany, FR Hong Kong Hungary Ireland Israel Italy	Argentina Australia Austria Belgium France Greece Hong Kong Hungary Ireland Israel Italy Japan Mexico	Argentina Austria Brazil Chile Greece Hong Kong Hungary Ireland Israel Italy Japan Mexico Peru	Argentina Brazil Chile Colombia Ecuador Greece Hong Kong Hungary Ireland Israel Italy Malaysia Mexico	Argentina Brazil Chile Colombia Ecuador Greece Hungary Ireland Israel Malaysia Mexico Poland Portugal

Developing countries 229

	Norway Poland South Africa Spain UK USSR Venezuela	Japan Mexico Netherlands Norway Poland South Africa Spain UK USSR Venezuela	Netherlands Norway Poland Portugal South Africa Spain UK USSR Venezuela Yugoslavia	Poland Portugal South Africa Spain UK USSR Venezuela Yugoslavia	Peru Poland Portugal South Africa South Korea Spain Taiwan USSR Venezuela Yugoslavia	South Africa South Korea Spain Taiwan Turkey USSR Venezuela Yugoslavia
1,200–2,300	Colombia Greece Guatemala Hong Kong Japan Mexico Peru Portugal Turkey Yugoslavia	Brazil Colombia Ecuador Greece Guatemala Malaysia Peru Portugal Turkey Yugoslavia	Bolivia Brazil Colombia Ecuador Guatemala Malaysia Peru South Korea Tunisia Turkey	Bolivia Colombia Ecuador Guatemala Malaysia Philippines South Korea Taiwan Thailand Tunisia Turkey	Bolivia Guatemala Morocco Philippines Thailand Tunisia Turkey	Guatemala Indonesia Morocco Peru Philippines Sri Lanka Thailand Tunisia
450–1,200	Bolivia Brazil Cameroon China Cote d'Ivoire Ecuador Egypt Ghana Guatemala Haiti	Bangladesh Bolivia Cameroon China Cote d'Ivoire Egypt Ghana Haiti India Indonesia	Bangladesh Cameroon China Cote d'Ivoire Egypt Ghana Haiti India Indonesia Kenya	Bangladesh Cameroon China Cote d'Ivoire Egypt Ghana Haiti India Indonesia Kenya	Bangladesh Cameroon China Cote d'Ivoire Egypt Haiti India Indonesia Kenya Madagascar	Bangladesh Bolivia Cameroon China Cote d'Ivoire Egypt Haiti India Kenya Madagascar

230 *Developing countries*

Table 8.12

	1960	1965	1970	1975	1980	1985
	India Indonesia Kenya Madagascar Malaysia Morocco Nigeria Pakistan Philippines Senegal South Korea Sri Lanka Taiwan Thailand Tunisia Zambia Zimbabwe	Kenya Madagascar Morocco Nigeria Pakistan Philippines Senegal South Korea Sri Lanka Taiwan Thailand Tunisia Zambia Zimbabwe	Madagascar Morocco Nigeria Pakistan Philippines Senegal Sri Lanka Taiwan Thailand Zambia Zimbabwe	Madagascar Nigeria Pakistan Senegal Sri Lanka Zambia Zimbabwe	Nigeria Pakistan Senegal Sri Lanka Zambia Zimbabwe	Nigeria Pakistan Senegal Zambia Zimbabwe
Below 450	Bangladesh Ethiopia Malawi Mali Tanzania Zaire	Ethiopia Malawi Mali Tanzania Zaire	Ethiopia Malawi Mali Tanzania Zaire	Ethiopia Malawi Mali Tanzania Zaire	Ethiopia Ghana Malawi Mali Tanzania Zaire	Ethiopia Ghana Malawi Mali Tanzania Zaire

Table 8.13 *Life expectancy, infant mortality and primary school enrolment (Source: 'The rich and the poor: changes in incomes of developing countries since 1960', Overseas Development Institute, London, 1988, Tables 3.4 and 5. Reproduced by permission of the Overseas Development Institute, London).*

Life Expectancy (years at birth)		Infant Mortality (deaths per 1,000)		Primary school enrolment (%, relevant age group)	
1960	1985	1960	1985	1960	1984
	70 and above		15 or fewer	Effectively universal	95 or above
Australia	Argentina		Australia		Argentina
Belgium	Australia		Austria	Argentina	Australia
Canada	Austria		Belgium	Australia	Austria
France	Belgium		Canada	Austria	Belgium
Netherlands	Canada		France	Belgium	Brazil
Norway	Chile		Germany, FR	Brazil	Cameroon
UK	France		Hong Kong	Canada	Canada
US	Germany, FR		Ireland	Chile	Chile
	Hong Kong		Israel	France	China
	Hungary		Italy	Germany, FR	Colombia
	Ireland		Japan	Greece	Ecuador
	Israel		Netherlands	Hungary	France
	Italy		Norway	Ireland	Germany, FR
	Japan		Spain	Israel	Greece
	Netherlands		UK	Italy	Hong Kong
	Norway		US	Japan	Hungary
	Poland			Malaysia	Indonesia
	Portugal		16–50	Netherlands	Ireland
	Spain	Australia	Argentina	Norway	Israel
	Taiwan	Austria	Chile	Philippines	Italy
	UK	Belgium	China	Poland	Japan
	US	Canada	Colombia	Portugal	Kenya
	USSR	France	Greece	Spain	Madagascar
	Venezuela	Germany, FR	Hungary	Sri Lanka	

232 *Developing countries*

Table 8.13

Life Expectancy (years at birth)		Infant Mortality (deaths per 1,000)		Primary school enrolment (%, relevant age group)	
1960	1985	1960	1985	1960	1984
	Yugoslavia	Greece	Malaysia	UK	Malaysia
		Hong Kong	Mexico	US	Mexico
	51–69	Hungary	Philippines	USSR	Netherlands
Argentina	Bangladesh	Ireland	Poland	Venezuela	Norway
Austria	Bolivia	Israel	Portugal	Yugoslavia	Peru
Brazil	Brazil	Italy	South Korea	Zimbabwe	Philippines
Chile	Cameroon	Japan	Sri Lanka		Poland
China	China	Netherlands	Thailand		Portugal
Colombia	Colombia	Norway	USSR		South Korea
Ecuador	Cote d'Ivoire	Spain	Venezuela		Spain
Germany, FR	Ecuador	UK	Yugoslavia		Sri Lanka
Greece	Egypt	US			Thailand
Hong Kong	Ghana	USSR			Tunisia
Hungary	Greece				Turkey
Ireland	Guatemala		51–99		UK
Israel	Haiti	Argentina	Brazil		US
Italy	India	Colombia	Cameroon		USSR
Japan	Indonesia	Guatemala	Ecuador		Venezuela
Malaysia	Kenya	Malaysia	Egypt		Yugoslavia
Mexico	Madagascar	Mexico	Ghana		Zaire
Poland	Malaysia	Poland	Guatemala		Zambia
Portugal	Mexico	Portugal	India		Zimbabwe
South Africa	Morocco	South Africa	Indonesia		
South Korea	Nigeria	South Korea	Kenya		51–94
Spain	Pakistan	Sri Lanka	Morocco	Bolivia	Bangladesh
Sri Lanka	Peru	Venezuela	Peru	Cameroon	Bolivia
Taiwan	Philippines	Yugoslavia	South Africa	Colombia	Cote d'Ivoire
USSR	South Africa		Tunisia	Ecuador	Egypt
Venezuela	South Korea		Turkey	Egypt	Ghana

					100 and above		less than 50	
Tanzania				Zimbabwe	Bangladesh	India		Haiti
Thailand			Bangladesh		Bolivia	Indonesia		India
Tunisia		Bangladesh	Bolivia		Cote d'Ivoire	Madagascar		Malawi
Turkey		Bolivia	Brazil		Ethiopia	Malawi		Morocco
Zaire		Cameroon	Cameroon		Haiti	Mexico		Nigeria
Zambia		Cote d'Ivoire	Chile		Madagascar	Peru		Senegal
Zimbabwe			China		Malawi	South Africa		Tanzania
			Cote d'Ivoire		Mali	South Korea		Zaire
50 and below			Ecuador		Nigeria	Thailand		
Ehiopia			Egypt		Pakistan	Tunisia	less than 50	
Malawi			Ethiopia		Senegal	Turkey	Ethiopia	
Mali			Ghana		Tanzania	Zaire	Mali	
Senegal			Haiti		Zaire		Pakistan	
			India					
			Indonesia				Bangladesh	
			Kenya				Cote d'Ivoire	
			Madagascar				Ethiopia	
			Malawi				Ghana	
			Mali				Guatemala	
			Morocco				Haiti	
			Nigeria				Kenya	
			Pakistan				Mali	
			Peru				Morocco	
			Philippines				Nigeria	
			Senegal				Pakistan	
			Tanzania				Senegal	
			Thailand				Tanzania	
			Tunisia				Zambia	
			Turkey					
			Zaire					
			Zambia					
			Zimbabwe					

50 and below
Bangladesh
Bolivia
Cameroon
Cote d'Ivoire
Egypt
Ethiopia
Ghana
Guatemala
Haiti
India
Indonesia
Kenya
Madagascar
Malawi
Mali
Morocco
Nigeria
Pakistan
Peru
Philippines
Senegal
Tanzania
Thailand
Tunisia
Turkey
Zaire
Zambia
Zimbabwe

8.2 Energy and development

8.2.1 *Current position*

Under presently accepted scenarios, developing countries will go on rapidly increasing their energy demand; e.g. over 30 years Africa increased commercial energy by 125%, Latin America by 150%, India by 175% and China by 150% [4], global energy meanwhile rising from 370 exajoules to 770 exajoules (Chapter 1). Developing countries, with 75% of world population, will consume 25% of world energy; per capita energy consumption in developing countries remaining a fraction of that in developed, e.g. Africa 0.3 tonnes of oil per capita in contrast to 5 tonnes per capita in Western Europe and more than 8 tonnes in North America. Western Europe transforms 150 times the quantity of fuels as Nepal. Historically, unlimited energy has been assumed necessary for development with no account taken of non-renewables, resource costs, and environmental maintenance. Currently developing country electricity needs capital of over US$ 125 billion per annum over the next 20 years. Since these countries already spend one quarter of public budgets on electricity, such huge capital expenditure will just not be available. Especially, the US$ 50 to 70 billion per annum foreign currency required over the same period is ten times more than that available, given debt service problems; this is one automatic way of limiting global environmental problems. However, in practice, the energy to income growth coefficients of energy and electricity, i.e. the percentage growths in energy or electricity needed to sustain a 1% growth in GDP, will decline with development (Chapters 1 and 2, see Figure 8.6), partly due to improving energy and electricity sector efficiency with development. Developing countries' energy losses are high at 20% in early development, again decreasing with development. Greater energy efficiency and conservation rapidly leads to less energy needs (Figure 8.6). Developing countries, like developed, unless special efforts are made, are subject to the undesirable, economically unfortunate effects of poor energy efficiency, little conservation and poor environmental maintenance. Figures 8.7 to 8.9 refer.

Figure 8.10 gives future hope, illustrating investment and fuel substitution effects, with Figures 8.11, 8.12 and 8.13 illustrating the precariousness and enormous complexity of the situation.

8.2.2 *Case study*

For the UK, examined at the developing stage [5], energy data is plentiful (Table 8.14, Figure 8.14). Energy grew at similar rates as GDP from 1700 to 1880, then faster, followed by periods of reversed trend; after World War II, energy continued to grow more slowly than GDP. Figure 8.15 shows energy consumption per capita, also per unit of output between 1700 and 1975. This has the profile of Figure 8.14, but shows energy per GDP falling slightly between 1800 and 1830, rising again to 1880, and falling thereafter; Figures 8.15, 8.16 and 8.17 showing GDP against energy, plus GDP per capita and energy consumption, both over time, confirm Figure 8.14 in the period 1830 to

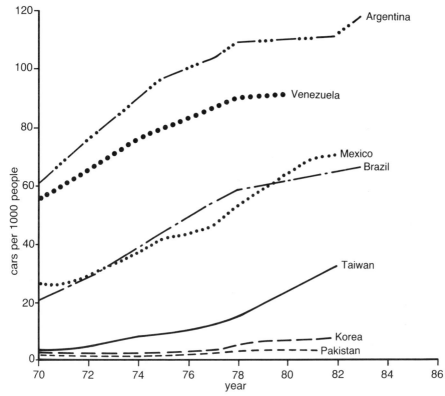

Figure 8.7 *Cars per capita (Source:* SCHIPPER, L. *'Energy demand in the Third World', 1986, Group Planning, Shell International Petroleum. Reproduced by permission of Shell International)*

1880, when energy consumption grew remarkably faster than GDP, perhaps because 1830 was a genuine turning point in energy growth *vis-a-vis* GDP, or because of merely an upsurge in coal consumption occurring at the time.

By 1800, between one-half and one-third of all UK coal produced was for households, including cottage industries. The conglomerated iron industry, the largest industrial energy consumer, took only 15% of the coal. Other large industrial users were: brickmakers, brewers, distillers, bakeries, potteries, copper and tin smelters. Only 2% of UK coal was exported. Table 8.15 shows coal development up to 1800, meaningfully considered alongside non-mineral fuels, i.e. wood, wind and water. Up to the sixteenth century, coal was considered technically and environmentally inferior to wood by industry and households, yet production and consumption grew rapidly from 1550 to 1700 because of timber shortages crises, causing rapid rises in fuelwood prices (Table 8.16). By 1700 timber burned as fuel was equivalent to about 500 tons of coal, while coal consumption was already three million. But in 1700 coke could not

Table 8.14 Population, output and energy consumption in the UK, 1700 to 1975 (Source: HUMPHREY, W.S. and STANISLAV, J. 'Economic growth and energy consumption in the UK, 1700 to 1975' Energy Policy, Vol. 1, No. 1, 1979, Table 1, p. 31, published by Butterworth-Heinemann, Ltd. Reproduced by permission of Energy Policy)

	Population		Output			Energy consumption (mineral fuels and hydro-power)				
	Million	Index (1800 = 100)	Index (1800 = 100)	Index of output/head (1800 = 100)	Million tce	Index (1800 = 100)	Consumption per head (tce/head)	Index of consumption per head (1800 = 100)	Index of consumption per unit output (1800 = 100)
1700	9.4	59	40	68	3	28	0.32	47	70
1760	11.0	69	59	85	5	46	0.45	66	78
1800	15.9	100	100	100	10.78	100	0.68	100	100
1810	18.1	113	122	108	13.73	127	0.76	112	104
1820	21.0	132	158	120	17.16	159	0.82	121	101
1830	24.1	153	226	149	21.86	203	0.91	153	90
1840	26.7	168	285	170	31.94	296	1.20	176	104
1850	27.4	172	339	197	45.78	425	1.67	246	125
1860	29.0	182	436	240	72.34	671	2.49	366	154
1870	31.6	199	534	268	97.64	906	3.09	454	170
1880	34.9	219	642	293	125.18	1161	3.59	528	181
1890	37.8	238	801	336	145.82	1353	3.86	568	169
1900	41.5	261	985	377	165.56	1536	3.99	587	156
1910	45.3	285	1104	387	188.56	1749	4.16	612	158
1920	43.4	273	1093	400	193.81	1798	4.12	606	158
1925	44.1	277	1191	430	182.60	1694	4.14	609	142
1930	44.8	282	1281	454	184.70	1713	4.12	606	134
1938	47.5	299	1531	512	196.58	1824	4.14	609	119
1950	50.6	318	1855	583	225.70	2094	4.46	656	113
1960	52.4	330	2415	731	264.90	2457	5.06	744	102
1965	54.4	342	2854	834	296.90	2754	5.46	803	96
1970	55.5	349	3196	916	328.00	3043	5.91	869	95
1975	56.4	355	3423	964	319.70	2965	5.67	834	87

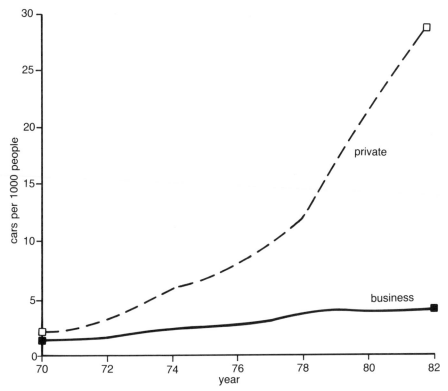

Figure 8.8 *Cars in Taiwan (Source:* SCHIPPER, L. *'Energy demand in the Third World', 1986, Group Planning, Shell International Petroleum. Reproduced by permission of Shell International)*

be substituted for wood in metal smelting and timber shortages persisted, iron smelting swallowing up enormous wood quantities.

The rapid growth of coal production from the mid-sixteenth to the late seventeenth century represents largely the substitution of coal for wood as household fuels. Eighteenth-century coal production continued to grow, especially later on because of population growth, drift into towns and demands of an expanding, changing economy. Coal consumption grew about the same rate as GDP between 1700 and 1800, 11 million tons being consumed in 1800. Even though wood was preferred to coal, by the 1700s coal was already a primary fuel used in aluminium manufacture, also copper, saltpetre and gunpowder, extraction of salt from salt water and from brine, boiling soap, refining sugar, making starch and candles, preparing foods and brewing. Part of coal consumption's growth involved substituting coal for wood in industry and the metal trades. Also, the late eighteenth century was the great canal building age which considerably reduced transport costs, lowering the coal price and increasing demand.

238 *Developing countries*

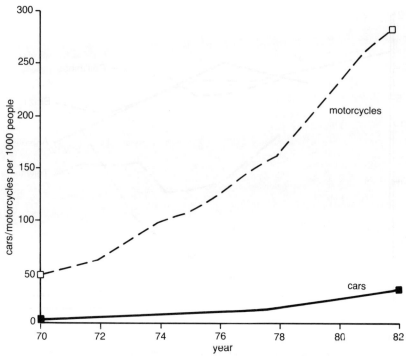

Figure 8.9 *Cars versus motorcycles, Taiwan (Source:* SCHIPPER, L. *'Energy demand in the Third World', 1986, Group Planning, Shell International Petroleum. Reproduced by permission of Shell International)*

Table 8.15 *Estimated annual production of the principal UK mining districts (tons) (Source:* HUMPHREY, W.S. and STANISLAV, J. *'Economic growth and energy consumption in the UK, 1700 to 1975' Energy Policy, Vol. 1, No. 1, 1979, Table 2, published by Butterworth-Heinemann, Ltd. Reproduced by permission of Energy Policy)*

Coalfield	1551–60	1681–90	1781–90
Durham and Northumberland	65 000	1225 000	3000 000
Scotland	40 000	475 000	1600 000
Wales	20 000	200 000	800 000
Midlands	65 000	850 000	4000 000
Cumberland	6 000	100 000	500 000
Kingswood Chase	6 000		140 000
		100 000	
Somerset	4 000		140 000
Forest of Dean	3 000	25 000	90 000
Devon and Ireland	1 000	7 000	25 000
Total	210 000	2982 000	10295 000

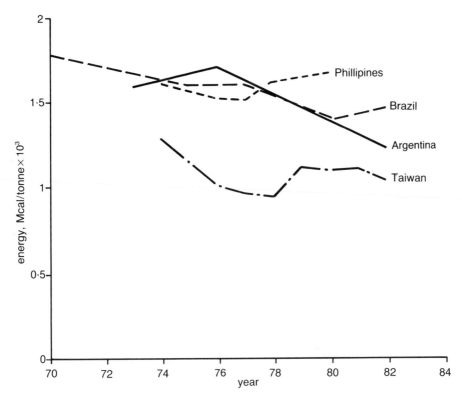

Figure 8.10 *Energy use in cement production (Source:* SCHIPPER, L. *'Energy demand in the Third World', 1986, Group Planning, Shell International Petroleum. Reproduced by permission of Shell International)*

Table 8.16 *Indices of estimated price movements in London 1451–1642 (Source:* HUMPHREY, W.S. and STANISLAV, J. *'Economic growth and energy consumption in the UK, 1700 to 1975' Energy Policy, Vol. 1, No. 1, 1979, Table 3, published by Butterworth-Heinemann, Ltd. Reproduced by permission of Energy Policy)*

Commodity	1451–1500	1531–40	1551–60	1583–92	1603–12	1613–22	1623–32	1633–42
General	100	105	132	198	251	257	282	291
Firewood	100	94	163	277	366	457	677	780
Coal	100	89	147	186	295	371	442	321

240 Developing countries

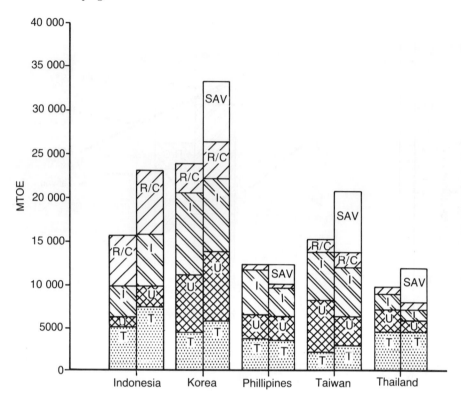

Figure 8.11 *Oil backout in Asia 1978–83 by sector*
R/C = residential/commercial
I = industry
U = utilities
T = transport
(Source: SCHIPPER, L. *'Energy demand in the Third World', 1986, Group Planning, Shell International Petroleum. Reproduced by permission of Shell International)*

8.2.3 *Energy elasticity*

It is important for developing countries to know what happens to the value of the energy and electricity elasticities with GDP growth as development progresses. Table 8.17 shows the energy coefficient for the UK from 1800 to 1975. There occurred a rise in mid-nineteenth century; it then fell. Data in Tables 8.14 to 8.17 indicate that energy and GDP grew at about the same rate during the 30 years from 1900.

Table 8.18 shows a dramatic turning point around 1830. The consumption of household fuels except coal rose with population to about 1830; GDP growth probably accelerated after 1830, partly due to using steam in manufacture (see Table 8.19). Another factor in UK economic growth was railway development.

Developing countries 241

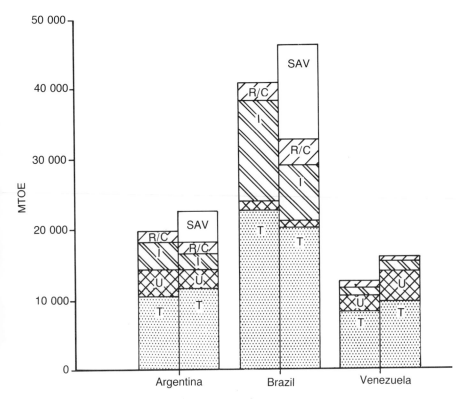

Figure 8.12 *Oil backout in Latin America, 1978–1983*
R/C = residential/commercial
I = industry
U = utilities
T = transport
(Source: SCHIPPER, L. *'Energy demand in the Third World'*, 1986, Group Planning, Shell International Petroleum. Reproduced by permission of Shell International)

The growth rates of mineral fuels consumption was mainly due to steam navigation coal (Table 8.20). Also, generally, railways mid-nineteenth century had a major change in the character of UK capital stock. Between 1880 and World War II, the UK energy elasticity fell sharply to about 0.62.

Table 8.21 shows the dominant position of iron and steel achieved by 1870. Figure 8.15 shows the years around 1880 were a peak for the elasticity, coinciding with the highest share the iron industry achieved in GDP (Table 8.22).

The period 1870 to 1914 saw improvements in efficiency in coal, also in other industries. It is difficult to infer anything about long-term trends from 1914 to 1950 due to two world wars. Between wars there was likely to be low energy

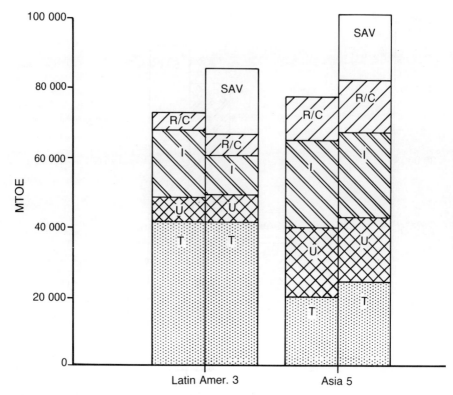

Figure 8.13 *Total oil backout, 1978 to 1983*
R/C = residential/commercial
I = industry
U = utilities
T = transport
(*Source:* SCHIPPER, L. 'Energy demand in the Third World', 1986, Group Planning, Shell International Petroleum. Reproduced by permission of Shell International)

elasticity; further changes took place away from energy-intensive industries plus continued improvements in energy efficiency, but with strikingly small flows of capital resources, mass unemployment, and depressed levels of GDP. The behaviour of the elasticity 1956 to 1975 is shown in Figure 8.18 which is a plot of five-year average elasticities against the year ending each five-year period.

8.2.4 *Lessons*

The UK experienced high energy elasticities during the period 1830 to 1870, with rapidly changing capital structure, paralleled in developing countries

Developing countries 243

Table 8.17 *Energy coefficients in the UK over the period 1800 to 1975 (Source: HUMPHREY, W.S. and STANISLAV, J. 'Economic growth and energy consumption in the UK, 1700 to 1975' Energy Policy, Vol. 1, No. 1, 1979, Table 4, published by Butterworth-Heinemann, Ltd. Reproduced by permission of Energy Policy)*

Period	Output growth (% pa)	Energy consumption growth (% pa)	Energy coefficient
1800–10	2·0	2·4	1·20
1810–20	2·6	2·3	0·89
1820–30	3·6	2·5	0·69
1830–40	2·4	4·0	1·67
1840–50	1·8	3·7	2·06
1850–60	2·5	4·7	1·88
1860–70	2·0	3·0	1·50
1870–80	1·9	2·5	1·32
1880–90	2·2	1·5	0·68
1890–1900	2·1	1·3	0·62
1900–10	1·2	1·3	1·08
1910–20	0·3	0·3	1·00
1925–30	1·5	1·2	0·80
1930–38	2·3	0·8	0·35
1938–50	1·6	1·2	0·75
1950–60	2·7	1·6	0·59
1960–70	2·8	2·2	0·79
1970–75	2·4	1·3	0·54

Table 8.18 *Comparison of growth rates and energy coefficients for the UK (including Eire) and Great Britain (excluding Eire) (Source: HUMPHREY, W.S. and STANISLAV, J. 'Economic growth and energy consumption in the UK, 1700 to 1975' Energy Policy, Vol. 1, No. 1, 1979, Table 5, published by Butterworth-Heinemann, Ltd. Reproduced by permission of Energy Policy)*

Period	Output growth (% pa) UK	GB	Energy consumption growth (% pa) UK	GB	Energy coefficient UK	GB
1830–40	2·31	2·77	4·0	4·5 (4·2)	1·67	1·63 (1·52)
1840–50	1·74	2·30	3·7	4·3 (4·6)	2·06	1·87 (2·00)
1850–60	2·54	2·76	4·7	4·9 (5·2)	1·88	1·78 (1·88)
1860–70	2·62	2·83	3·0	3·1 (3·4)	1·50	1·10 (1·20)
1870–80	2·00	2·77	2·5	2·6 (2·8)	1·32	0·94 (1·01)
1880–90	1·91	2·03	1·5	1·6 (1·8)	0·68	0·79 (0·89)
1890–1900	2·39	2·47	1·3	1·3 (1·5)	0·62	0·53 (0·61)
1900–10	1·19	1·20	1·3	1·3 (1·4)	1·08	1·08 (1·17)

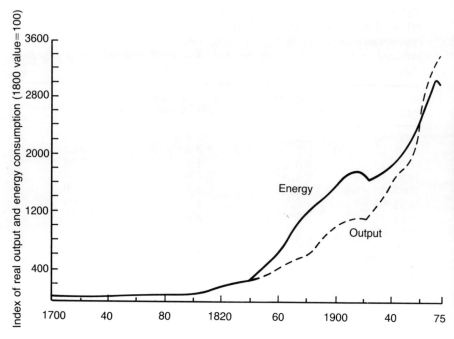

Figure 8.14 *Indices of real output and energy consumption for the UK 1700–1975 (Source:* HUMPHREY, W.S. and STANISLAV, J. *'Economic growth and energy consumption in the UK, 1700 to 1975' Energy Policy, Vol. 1, No. 1, 1979, Figure 1, p. 31, published by Butterworth-Heinemann, Ltd. Reproduced by permission of Energy Policy)*

Table 8.19 *Motive horse power in textile factories in the UK (Source:* HUMPHREY, W.S. and STANISLAV, J. *'Economic growth and energy consumption in the UK, 1700 to 1975' Energy Policy, Vol. 1, No. 1, 1979, Table 6, published by Butterworth-Heinemann, Ltd. Reproduced by permission of Energy Policy)*

Year	Water	Steam	Total
1839	20 418	62 846	83 264
1850	26 104	108 113	134 217
1856	23 724	137 713	161 437
1861	29 359	375 311	404 670
1868	29 830	337 851	367 681
1870	27 321	478 434	505 755

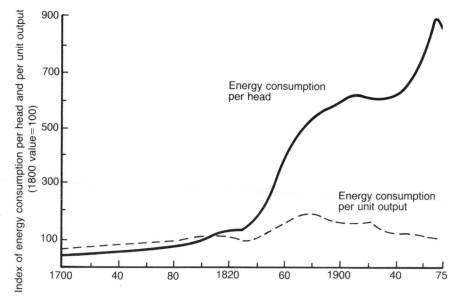

Figure 8.15 *Indices of energy consumption per head and energy consumption per unit output for UK, 1700 to 1975 (Source: HUMPHREY, W.S. and STANISLAV, J. 'Economic growth and energy consumption in the UK, 1700 to 1975' Energy Policy, Vol. 1, No. 1, 1979, Figure 2, published by Butterworth-Heinemann, Ltd. Reproduced by permission of Energy Policy)*

Table 8.20 *UK mineral fuels consumption growth 1830 to 1890 with and without steam navigation coal consumption (Source: HUMPHREY, W.S. and STANISLAV, J. 'Economic growth and energy consumption in the UK, 1700 to 1975' Energy Policy, Vol. 1, No. 1, 1979, Table 7, published by Butterworth-Heinemann, Ltd. Reproduced by permission of Energy Policy)*

	Mineral fuel consumption growth rate (% pa)	
Period	Including coal for steam navigation	Excluding coal for steam navigation
1830–1840	3·9	3·7
1840–1870	3·8	3·6
1870–1890	2·0	1·5
1830–1870	3·8	3·6
1830–1890	3·2	2·9

246 *Developing countries*

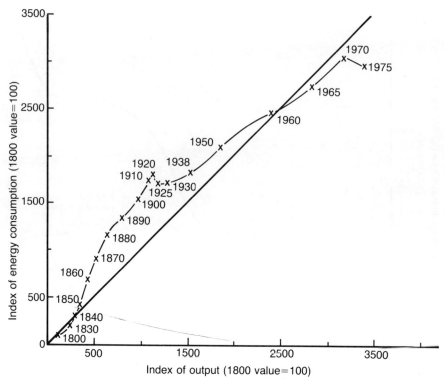

Figure 8.16 *Energy consumption versus output (Source: HUMPHREY, W.S. and STANISLAV, J. 'Economic growth and energy consumption in the UK, 1700 to 1975' Energy Policy, Vol. 1, No. 1, 1979, Figure 3, published by Butterworth-Heinemann, Ltd. Reproduced by permission of Energy Policy)*

Table 8.21 *Estimation of distribution of UK coal consumption in various years (Source: HUMPHREY, W.S. and STANISLAV, J. 'Economic growth and energy consumption in the UK, 1700 to 1975' Energy Policy, Vol. 1, No. 1, 1979, Table 8, published by Butterworth-Heinemann, Ltd. Reproduced by permission of Energy Policy)*

	1840	1869	1887	1913	1929
Iron industry	26	33	19	16	14
Mines	3	7	8	10	7
Steam navigation	2	6	15	9	9
Gas and electricity	2	7	7	12	16
General manufacturing	34	28	31	33	32
Domestic	33	19	20	20	22
Total	100	100	100	100	100

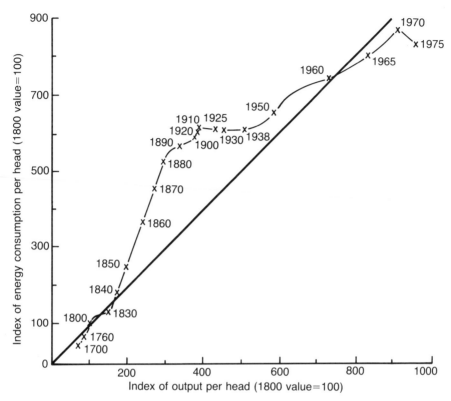

Figure 8.17 *Energy consumption per head versus output per head for the UK (Source: HUMPHREY, W.S. and STANISLAV, J. 'Economic growth and energy consumption in the UK, 1700 to 1975' Energy Policy, Vol. 1, No. 1, 1979, Figure 4, published by Butterworth-Heinemann, Ltd. Reproduced by permission of Energy Policy)*

today, but:

(a) Developing countries today benefit from a much higher level of energy efficiency than was available to the UK in the nineteenth century
(b) Domestic fixed investment as a proportion of GDP and rates of structural changes in developing countries are much more rapid than anything in UK economic history
(c) Energy growth, measured by mineral fuels consumption and hydro power, inflates underlying trends today compared with nineteenth century UK, because wood, peat, animal and other renewables have continued to be of considerable importance in all today's developing world.

From 1960 to 1972 OECD overall had an energy elasticity 1.02, whereas non-communist non-OPEC developing countries averaged 1.26, East Asian countries, undergoing rapid restructuring, having the highest elasticity of 1.59,

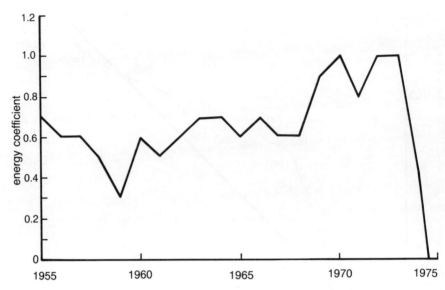

Figure 8.18 *Energy coefficient, 1955–1975 (Source:* HUMPHREY, W.S. and STANISLAV, J. *'Economic growth and energy consumption in the UK, 1700 to 1975' Energy Policy, Vol. 1, No. 1, 1979, Figure 5, published by Butterworth-Heinemann, Ltd. Reproduced by permission of Energy Policy)*

Table 8.22 *Approximate estimates of the Iron Industry's share in the National Product of Great Britain, 1805–1907 (Source:* HUMPHREY, W.S. and STANISLAV, J. *'Economic growth and energy consumption in the UK, 1700 to 1975' Energy Policy, Vol. 1, No. 1, 1979, Table 9, published by Butterworth-Heinemann, Ltd. Reproduced by permission of Energy Policy)*

Year	Gross output (less coal and imported ore) as a percentage of GNP
1805	5·9
1818	3·4
1821	3·6
1831	3·6
1841	3·8
1851	6·2
1861	7·6
1871	11·6
1881	10·3
1891	7·5
1901	5·8
1907	6·4

with South Asia, dominated by India, at 1.69. The figure for Central and South America overall was 1.19. Evidently, periods of industrialisation involving rapid structural economic change end in the capital stock needing higher growth in energy consumption than in GDP.

A brief mention needs making here of the allied subject of traditional energy being replaced by commercial in developing countries as they were in the UK. As mentioned earlier, fuel wood and charcoal still continue today to be the most widely used fuels in most developing countries because, despite growing scarcity, they are much cheaper than any alternative, especially including appliances' costs. One alternative, electricity, is ruled out while people live either in remote villages or slum town dwellings which are both unlikely to be electrified. Kerosene and electricity, both widely used for lighting, are not used very much for cooking. Electricity will therefore not take over from traditional fuels within the foreseeable future.

8.3 Electricity and development

8.3.1 *Electricity's importance*

Even in the least developed countries, reliable electricity at a reasonable price is important for economic growth. Until recently, the critical issue has always been seen as electricity demand exceeding supply, constrained by financial, technical, and manpower shortages. Demand management techniques, improved energy efficiency, conservation and dynamic pricing now enable demand to be kept much more in step with supply; Chapters 3 and 4 showed that these measures were viable economic and financial alternatives to providing new generating, transmission, distribution and utilisation plant. In developing countries, because of the capital intensity of electricity, it is vital in any case to save on as many new high capital cost schemes as possible. The World Bank estimates [6] as US$ 1 trillion the investment cost for electricity in the 1990s for 70 developing countries, based on 1990 country expansion plans for the next ten years, for a 6·6% average annual growth in electricity and projected costs for generation, transmission and distribution equipment. Allowing for realistic efficiency improvements and conservation, generation is expected to increase by 82%, most coming from coal, 40%, and hydro 38%, oil accounting for 10%, gas 8%, nuclear 3%, and geothermal, 1%. Nearly half of the investment will be in India and China.

Developing countries in the 1990s are actively seeking increased technical, financial, economic and managerial efficiency in their electricity sectors because any savings possible are likely to be immense (Tables 8.23 to 8.29). Developing countries have made significant improvements in their electricity sectors over the last decades and such sectors enter the 1990s generally with better organisations than other sectors. However, during times of high growth, utilities weathering oil price increases and high inflation hampered financial targets, governments then being slow to grant tariff increases. Urgent needs thus often arise to arrest deteriorating investment trends, given that electricity investments sometimes absorb nearly half of all public investments, often causing severe debt problems for governments. In the 1970s, the decline in overall

Table 8.23 *Comparison of 1989 and 1999 LDC installed capacity (Source:* MOORE, E. *'Electric power in developing countries in the 1990s', Industry and Energy Department, World Bank, Table 1, 1990. Reproduced by permission of the World Bank)*

Type	1989 GW	1989 %	1999 GW	1999 %
Hydro	185	39.3	322	37.7
Geothermal	2	0.4	5	0.6
Nuclear	14	3.0	38	4.4
Oil thermal[1]	70	14.8	84	9.7
Gas thermal[2]	31	6.7	65	7.8
Coal thermal	169	35.8	341	39.8
Total	471	100.0	855	100.0

[1] Includes steam, combustion turbine, combined cycle and diesel
[2] Includes steam, combustion turbine and combined cycle

Table 8.24 *1989–1999 capacity additions in the LDCs (Source:* MOORE, E. *'Electric power in developing countries in the 1990s', Industry and Energy Department, World Bank, Table 2, 1990. Reproduced by permission of the World Bank)*

Plant type	GW	%
Hydro	137	35.7
Geothermal	3	0.8
Nuclear	24	6.2
Oil thermal – steam	10	2.7
– combustion turbine	1	0.3
– combined cycle	1	0.3
– diesel	1	0.2
– subtotal	13	3.5
Gas thermal – steam	15	4.4
– combustion turbine	3	0.9
– combined cycle	16	4.0
– subtotal	34	9.3
Coal thermal	172	44.5
Total	384	100.0

Table 8.25 *1989 and 1999 electricity supply in the LDCs (Source:* MOORE, E. *'Electric power in developing countries in the 1990s', Industry and Energy Department, World Bank, Table 3, 1990. Reproduced by permission of the World Bank)*

	1989		1999	
	TWh	%	TWh	%
Hydro	674	33·2	1207	31·5
Geothermal	11	0·6	29	0·8
Nuclear	80	3·9	212	5·5
Oil thermal	224	11·0	255	6·6
Gas thermal	120	5·9	332	8·6
Coal thermal	907	44·7	1793	46·6
Net imports	14	0·7	16	0·4
Total	2030	100·0	3844	100·0

utility performance for these and other reasons was paralleled by moves towards large monolithic, government controlled and funded utilities. This was because the universal rapid growth of electricity, increasing economies of scale, improved coordination and efficiency, greater reliability of electricity supply, all needed large, long-term investments. Thus there was a general move towards nationalisation and elimination of foreign ownership of utilities. In developing countries, many rationales for continuing concentration of electricity resources

Table 8.26 *1989 to 1999 increase in LDC annual electricity supply (Source:* MOORE, E. *'Electric power in developing countries in the 1990s', Industry and Energy Department, World Bank, Table 4, 1990. Reproduced by permission of the World Bank)*

	TWh	%
Hydro	533	29·4
Geothermal	18	1·0
Nuclear	132	7·3
Oil thermal	31	1·7
Gas thermal	212	11·7
Coal thermal	886	48·8
Net Imports	3	0·1
Total	1815	100·0

Table 8.27 Regional breakdown of LDC capacity expected to be added in the 1990s (Source: MOORE, E. 'Electric power in developing countries in the 1990s', Industry and Energy Department, World Bank, Table 5, 1990. Reproduced by permission of the World Bank)

Plant Type	GW				
	Asia	EMENA	LAC	Africa	Total
Hydro	78	15	42	2	137
Geothermal	2	—	1	—	3
Nuclear	14	7	3	—	24
Oil thermal	−3	4	12	1	14
Gas thermal	13	18	2	1	34
Coal thermal	140	25	6	1	172
Total (GW)	244	69	66	5	384
(%)	63·5	18·0	17·2	1·3	100·0

still remain valid, despite a growing worldwide consensus for improving utility efficiency and organisational reform of the sector and competition:

(a) In developing countries, electricity is vital for growth, industrial modernisation only being sustainable by large investments, and these to be efficient for the whole economy must be centrally directed.
(b) Utilities are a well recognised tool for helping with socio-economic problems, e.g. employment and income redistribution.
(c) Electricity has always proved to be a good vehicle for raising resources and taxing away surpluses.

Table 8.28 LDC power, capital expenditure in the 1990s (Source: MOORE, E. 'Electric power in developing countries in the 1990s', Industry and Energy Department, World Bank, Table 6, 1990. Reproduced by permission of the World Bank)

	1989$ Billions	%
Generation	448	60·0
Transmission	81	10·9
Distribution	152	20·5
General	64	8·6
Total	745	100·0

Table 8.29 *Regional breakdown of LDC power capital expenditure in the 1990s (Source:* MOORE, E. *'Electric power in developing countries in the 1990s', Industry and Energy Department, World Bank, Table 7, 1990. Reproduced by permission of the World Bank)*

	1989 $ Billions					
	Asia	EMENA	LAC	Africa	Total	%
Generation	277	82	83	6	448	60·0
Transmission	39	8	32	2	81	10·9
Distribution	100	23	27	2	152	20·5
General	39	11	13	1	64	8·6
Total	455	124	155	11	745	100·0
%	61·1	16·6	20·8	1·5	100·0	

However, while electricity has brought many benefits, scarcities of all resources focus attention on sector weaknesses and the need for it to improve resource efficiency. Past failures between governments, utilities, and consumers have been often due, not to policy shortcomings, but to imprecise interpretation and application of what were basically sound objectives. As a result, desire for growth and modernisation resulted in unquestioned funding of electricity, continuous government subsidies, non-optional and unbalanced investments, plus lack of productive efficiency and incentives to maintain technical and financial discipline. Social equity and employment objectives often led to excessive consumers' subsidies, inefficient pricing, and inadequate resource mobilisation. Attempts to tax away sector surpluses created highly skewed price structures, cross-subsidies and incorrect price signals. All this encouraged counter-productive government intervention. These considerations and desperate utility circumstances called for new approaches, particularly decentralisation and private participation in the sector (Chapter 5). Especially in developing countries, private participation is only likely to be successful if it is one element in a much broader economic package, involving policy reforms in trading, industry, and finance, with private capital infusion, transfers to private ownership and guarded introduction of entrepreneurship. However, ownership or management transfer to private hands will not solve all electricity sector problems, it is but one of several approaches to try to improve utilities' efficiency.

Any developing country utility reform must recognise that:

(a) Utility internal conditions must provide incentives for technical, financial, and managerial efficiency
(b) Remedies must try to address the whole range of problems plaguing utilities in the 1990s: weak planning, inefficient operation, inadequate maintenance, high technical and non-technical losses, low supply quality

(c) There are frequent blackouts, price freezes, change of management, excessive staffing, low salaries, poor morale and performance, undue government interference, such problems having persisted over decades, despite many consultant studies and institutional programmes

(d) Utility environment must in general be restructured including: delegation of authority, greater accountability, rational regulation, government control only at arm's-length, plus reforms in the financial climate and better access to capital

(e) Decisions must allow for uncertainty with respect to all the main parameters: capital cost, demand, inflation, exchange rates, interest rates and fuel prices, to be catered for by scenarios not a deterministic approach.

8.3.2 Electricity's position

Despite deteriorating performance, access to electricity in developing countries has increased considerably in both average per capita kWh generation and the percentage of population served. In 51 World Bank surveyed countries [7] generation increased at an annual average rate of 7%, from 196 kWh per capita in 1968 to 529 kWh in 1982. In that period, the average rate of per capita GDP growth was only 2%. The annual average growth rate of connections for 29 electricity projects was 9%. Even countries with little or negative economic growth had 5% annual growth in generation and 3% annual growth in connection. This required sustained high rates of investment and expansion in total assets of up to 15% annually in many countries. These investment needs create a great economic burden, many utilities often being forced to scale down programmes. Furthermore, investments have recently been weighted towards generation, with underfunding of transmission but especially distribution. One consequence of this is high electrical losses in many developing countries (Chapter 4) driving up costs and increasing financial burdens. They also indicate poor supply quality. Frequently even small distribution investments can achieve significant energy loss savings. There are also considerable non-electrical losses in developing systems, due to theft plus inadequate metering, billing, and collection. For the above-mentioned World Bank reviewed countries, total network losses during 1967–78 averaged 13%; reasonable norms would be less than 8%. Unfortunately, in most cases, losses and supply quality have progressively got worse or remained unchanged, despite large investments in the sector as a whole.

Developing country tariffs have lagged behind costs; the average operation ratio, total costs as a fraction of total revenue, for over 300 projects reviewed by the World Bank, deteriorated from 0·65 in 1966–73 to 0·80 during 1980–85. Worse are declining trends for overall returns on investments, from 10% in the early 1960s to 5% by 1985 (Figure 8.19). Despite increasing emphasis by lending agencies on economic efficiency pricing, only modest progress in this matter has been achieved. This is related to the reasons for persistently overoptimistic demand forecasting (Chapter 3), often attributable to frequent slowdowns in construction. Another aspect, institutional performance (e.g. the number of consumers per employee, the adequacy of maintenance and general

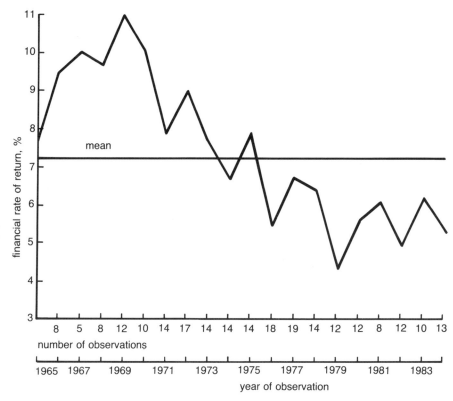

Figure 8.19 *Long-term decline in rates of return on investment (Source: MUNASINGHE, M. 'Power for development', IEE Review, March 1989, Figure 1. Reproduced by permission of the IEE Review)*

utility efficiency) stagnated to a relatively unsatisfactory level over the 1980s, attributable to high demand growth, frequently calling for plant doubling every seven to ten years. Labour use efficiency is indicated by numbers of consumers per employee. Excluding wholesale users, two-thirds of utilities surveyed had fewer than 100 connections per employee and only 10% had more than 150; these rates are about half what they perhaps could be. Maintenance is more difficult to measure quantitatively. Surveys indicate almost half the utilities fell into the poor to very poor maintenance category. Qualitative assessment of general utility efficiency has indicated similarly weak results.

8.3.3 Prospects

Thus, needs for greater electricity efficiency arise largely from large investment burdens and poor actual performance, these being especially serious for countries with serious debt problems. Electricity accounting for three-quarters of total energy investment, on average developing countries' energy investments

Table 8.30 *Estimated annual energy investment as a proportion of annual total public investment during the early 1980s (Source: MUNASINGHE, M. 'Power for development', IEE Review, March 1989, Table 1. Reproduced by permission of the IEE Review)*

Over 40%	30–40%	20–30%	10–20%	0–10%
Argentina	Ecuador	Botswana	Benin	Ethiopia
Brazil	India	China	Egypt	
Colombia	Pakistan	Costa Rica	Ghana	
Korea	Philippines	Liberia	Jamaica	
Mexico	Turkey	Nepal	Morocco	
			Nigeria	
			Sudan	

still account for 25% of total public capital (Table 8.30), while oil-importing countries are spending an average of 15% of export earnings on petroleum imports, while fuelwood shortages and deforestation problems continue unabated. High electricity energy related expenditures have contributed significantly to the indebtedness problems. Figure 8.20 indicates the alarming upward

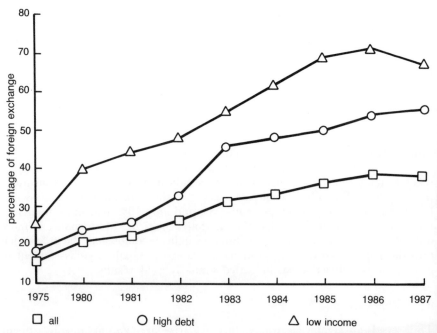

Figure 8.20 *Developing countries debts (Source: MUNASINGHE, M. 'Power for development', IEE Review, March 1989, Figure 2. Reproduced by permission of the IEE Review)*

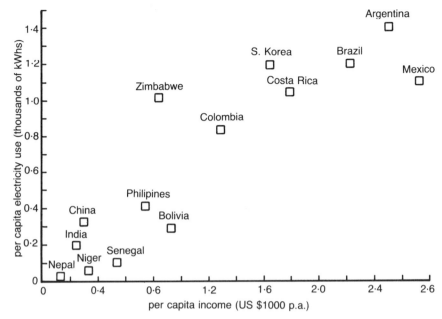

Figure 8.21 *Relationship between per capita income and power use (Source: MUNASINGHE, M. 'Power for development', IEE Review, March 1989, Figure 3. Reproduced by permission of the IEE Review)*

trend of developing countries' debts as a fraction of GDP from 1975 to 1987.

Yet electricity has a vital role to play in development. Figure 8.21 indicates this relationship. World Bank and UN data indicates the ratio of growth rates of electricity capacity and GDP is about 1.4 in developing countries. World Bank projections of economic growth up to 1995 are 4·2% annually for a base or medium growth scenario, and 5·6% for a high growth scenario. Assuming the same values of elasticity between electricity and GDP growth rates, installed power capacity is expected to grow at about 5·9% and 7·8% annually for the medium and high growth scenarios respectively, implying developing countries requiring additional electricity capacity ranging from 430 to 630 GW during 1989 to 1998, starting from 560 GW in 1988. Corresponding investment needs are large; the average cost of installed capacity being (say) US$ 1900 per kW (1988 figures) and thus, even under medium economic growth the required investment over 1989–98 is US$ 820 billion, an average US$ 82 billion annually. The financing requirement for high growth would be 50% greater. In comparison, present developing countries' annual investment rates are only US$ 50 billion, already difficult to maintain. To avoid this impasse, developing countries must urgently seek out and implement effective remedial demand and supply side issues and options, e.g. demand management and dynamic pricing.

8.4 Government role

8.4.1 *Required roles*

Despite recent emphasis on electricity deregulation, in developing countries governments must always play a special role because:

(a) Extreme capital shortage, local and foreign, in developing countries means publicly and privately owned utilities need government assistance, e.g. to obtain or guarantee loans on reasonable terms for as much capital as possible, from aid agencies and commercial sources

(b) Capital intensity of electricity from the nation's and the utility's points of view means no capital should ever be wasted, e.g. by excessive or poor supply standards, inappropriate technology, cost overruns, especially on large civil works, glossing over past mistakes

(c) Fuels are often a major issue which means vital questions arise, these best addressed jointly by government and utilities, but often needing direct government intervention, e.g. when exporting indigenous fuels or using them locally, for local pricing of such, their depletion rate, estimating resource cost of these fuels, likely medium and long-term comparative prices of alternative fuels which can be imported, say coal, oil, or natural gas, determining which fuels are presently being used by private and public industry and the electricity sector, and which fuel should be used for these purposes in the future.

(d) Having a national energy policy means that right signals should be given to consumers at all times, market forces not being good enough. Some government role is needed in developing countries

(e) A standard methodology for comparing all alternative projects is necessary across utilities, giving a consistency at least internally within a particular country, to ensure not only that optimum decisions are made but that they are seen to be made by all involved; this matter best being seen to by government.

8.4.2 *Sector adjustment*

The early 1980s ushered in a period likely to intensify during the 1990s, especially in developing countries, of structurally adjusting economies and sectors to more exactly suit world and national conditions in the 1990s with all the changes in emphasis already mentioned in other chapters, e.g. towards privatisation and deregulation of utilities. Governments play a vital role in this. Energy sectors, especially electricity, are prime targets for such restructuring because of their high capital-intensity and major use of other resources, e.g. fuels. Many developing country governments initiate their own sector adjustment, many are assisted by the World Bank. Government's role is first to make recommendation on how the investment programme for (say) electricity can be tailored to fit available resou, ces, at least cost, for short, medium and long-term economic growth; second, and of equal importance, to evaluate systems of investment programming by utilities to identify ways in which this can be improved, to provide a basis for better decisions to be taken in the future. Private utilities need government participation at least in some measure, e.g.

through regulatory authorities set up by government to periodically review and adjust national electricity in general, electricity markets and a utility's share in the sector in particular. In principle, where utilities are using public funds or relying on government guarantees for their borrowing, there is a legitimate argument for comparing their use of funds with the returns to other uses. However, although government as the main shareholder should address financial viability periodically, it also has socio-economic responsibilities, e.g. environmental maintenance, resource conservation, and it should in any case leave how to attain financial viability day-by-day to utility management.

8.5 Special issues and options

8.5.1 *Pricing*

Most issues and options special to developing countries are mentioned in the relevant chapters. World Bank case studies highlight the following practical issues for special consideration in electricity pricing:

(a) Distribution tariff distortion due to long-run marginal costs greatly exceeding existing bulk supply kW charges
(b) Unduly favourable treatment of some consumer classes, usually residential
(c) Need to increase or abolish block tariffs for residential consumers
(d) Distortions in the electricity and other energy marketplaces due to domestic oil, gas, and coal prices departing seriously from their opportunity costs
(e) Need to introduce dynamic pricing (Chapter 6)

Of particular importance for the 1990s is the estimation of LMC of electricity under two conditions: (a) continued technical progress, leading to steady decline in both investment costs and running costs; and (b) marked divergence between economic and financial costs, e.g. the social costs of low-grade coal is often lower than its financial cost, while for foreign exchange reasons the reverse is true.

8.5.2 *Efficiency, conservation and the environment*

In order to lessen the difficulties to developing countries (at least in the short-term) of taking care of these items properly, two major things are required, mainly from developed countries and the multi-lateral financing agencies, such as the World Bank and cofinanciers:

(1) A new socio-economic-financial methodology for assessing satisfactorily short-term viability which does not depart from medium and long-term economic viability, such a methodology including such well known economic and financial techniques as tilted annuities, variable discount rates, convex shadow pricing, and modern ways of coping with risk and uncertainty (Chapter 3)

260 *Developing countries*

(2) A new mix of lending by all financing sources, public, official and private, moving towards grants and soft loans, with variable and renegotiable servicing conditions, for projects including either a large proportion for technology transfer or entirely for technology transfer, projects aimed at demand management, improving efficiency, conservation and environmental maintenance, it being realised that worldwide projects for these purposes would in the medium and long-term more than recover the deficit between their receipts from normal lending and this new concessionary lending.

8.6 Special financing

8.6.1 *Financial engineering*

As mentioned earlier, electricity growth in the past has been prodigious in developing countries (Tables 8.23 to 8.29). Detailed utility financing plan requirements depend on the makeup of the development programme projects, e.g. proportion of capital to running costs, foreign to local currency and capital to labour for maintenance. Financial engineering fits the best finance in loan conditions, e.g. interest repayment period, etc., to these detailed characteristics of individual projects or development programmes. Projects requiring much capital in time to plan and construct, require different finance from projects needing less capital and a short lead period, e.g. gas turbines or diesels. Figures 8.22 through to 8.26 indicate the equipment type expected to be installed in developing countries in the 1990s and for which sound financing engineering is essential to utilities, borrowers and lenders alike.

Developing country electricity financing problems in the 1990s are likely to be [8]:

(a) The world economy is still passing through a variable period on oil and commodity prices, and also growth, which is likely to last through the 1990s.

(b) Oil importing developing countries need substantially larger external capital flows in loans and direct foreign investments to sustain acceptable economic growth rates, representing about 2·5% to 3·5% of GDP compared with past levels of 1·5% to 2·0% before 1973; proposals need continuous discussion to make these large flows possible, with mixed success likely.

(c) Traditionally, flows from official bilateral and multilateral aid have been about equal to those from private lenders and investors; although desirable to greatly increase the proportion of official flows, these being on much better loan terms than private ones, much of the needed additional funds just have to be obtained from private sources, particularly international (Eurocurrency) markets; although few developing countries could borrow in the past from private markets under favourable conditions, e.g. in Eurobonds, many more must be allowed to do so in the 1990s.

(d) More specifically, continuous adjustment to imported oil prices means a major reallocation of investments in developing nations, towards indige-

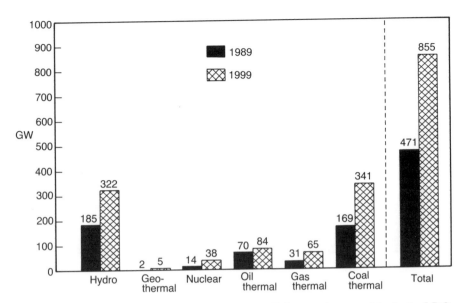

Figure 8.22 *Comparison of 1989 and 1999 installed generating capacities in the LDCs (Source: MOORE, E. 'Electric power in developing countries in the 1990s', Industry and Energy Department, World Bank, Figure 1. Reproduced by permission of the World Bank)*

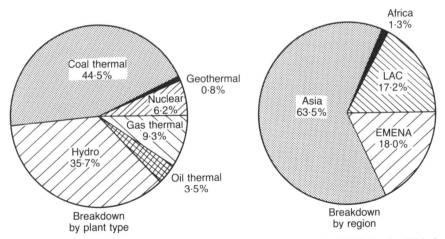

Figure 8.23 *Breakdown of capacity expected to be added in the LDCs in the 1990s by plant type and region (Source: MOORE, E. 'Electric power in developing countries in the 1990s', Industry and Energy Department, World Bank, Figure 2. Reproduced by permission of the World Bank)*

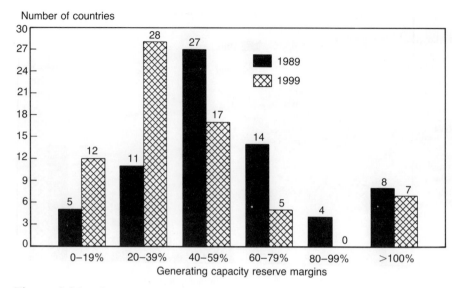

Figure 8.24 *Comparative generating capacity reserve margins in 69 LDCs for 1989 and 1999 (Source:* MOORE, E. *'Electric power in developing countries in the 1990s', Industry and Energy Department, World Bank, Figure 3. Reproduced by permission of the World Bank)*

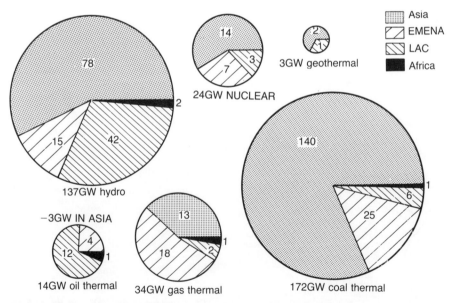

Figure 8.25 *Breakdown of generating capacity expected to be added in the 1990s by plant type (Source:* MOORE, E. *'Electric power in developing countries in the 1990s', Industry and Energy Department, World Bank, Figure 4. Reproduced by permission of the World Bank)*

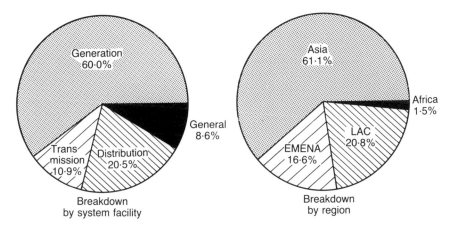

Figure 8.26 *Breakdown of capital expenditures for power in the LDCs in the 1990s by system facility and region (Source: MOORE, E. 'Electric power in developing countries in the 1990s', Industry and Energy Department, World Bank, Figure 5. Reproduced by permission of the World Bank)*

nous fuels and non-fuel burning generation; by 1985 such investments took twice the percentage of GDP and savings allocated to them in the early 1970s.

(e) The major shift in allocation of investments to electricity will not initially appear as higher proportions of external borrowing for this purpose, but only because of the large increases in total which will take place during the period of continuous economic adjustments to efficiency. However, when total external capital flows return to normal, at 1·5%–2% of GDP, the shift of foreign borrowing towards electricity will become evident, increasing from 13% to about 25% of total flows by the mid-1990s.

(f) About 60% of past borrowings for electricity came from official sources at lower interest rates and longer repayment periods than private loans; given the 1980s trends in official funds the percentage will reverse for the 1990s, 60% coming from private source; World Bank policy supporting this.

(g) In the short and medium term, energy crises will not be ones of shortages of resources or technology, only financial, especially in local currency.

8.6.2 Investments

Although energy to GDP elasticities have tended to approach unity, historically electricity growth has been about 1.5 times the energy and GDP growths. For example, during the 1960s electricity growth averaged 9%–10% in developing countries, with economic growth at 5% to 6%. Financial requirements for electricity therefore increase as a proportion of GDP, this tendency being compensated by technological developments and economies of scale which

Table 8.31 Shift in generation mix (Source: World Bank published studies. Reproduced by permission of the World Bank)

	Hydro	Oil	Coal	Other
Group 1				
1974 (%)	21	63	15	1
1990 (%)	12	18	10	60
	Hydro	Oil/Gas	Coal	Other
Group II				
1971 (%)	36	29	35	—
1987 (%)	25	14	30	31

match the electricity growth, by larger per unit costs for generators and networks. The net result was the 1960–1980 electricity expansion requirements staying on average at about 7%–9% of gross fixed capital formation. During the period 1980–2000 electricity and GDP growths are likely to remain about the same, both possibly declining somewhat. World Bank studies for two different country groups point to major shifts in generation mix with a sharp decrease in oil burning plants and increases in lignite, gas turbines, combined cycle and diesel plants (Table 8.31).

Results of these factors are: continued faster than GDP growth of electricity demand; less scope for economies of scale; faster than average price increases in equipment and heavy construction; and, particularly, higher capital cost of all plant, the proportion of developing countries' national investment taken by electricity projected to increase by about one-half, from about 7%–8% in 1970–1980 to about 10%–12% in the period 1990–2000. Higher fuel prices do not result only in bigger electricity investments but mean also more investment in exploration, developing other domestic energy sources. In OECD countries, annual energy investments' share of GDP is forecasted to increase from about 7·7% in the early 70s to about 18% by 2000. Of these investments, exceeding US$ one trillion (mentioned earlier), about one half are for electricity. Similar increases will take place in countries where investments in indigenous energy will require an additional US$40–80 billion to 2000. Although only 15% to 25% of primary energy goes into electricity, the latter requires 60% to 80% of all energy investments because of its high capital intensiveness.

8.6.3 Financing sources

World Bank studies over the last ten to fifteen years show:

(a) Foreign capital flows provided 70%–85% of foreign exchange needed for electricity; however, low income countries possibly received considerably less, about 35%–45%, while mid income countries received 70%–80%, and high income countries received all they needed.

Table 8.32 *Sources of electricity loans in developing countries (Source:* World Bank *published figures. Reproduced by permission of the World Bank)*

	%
Official sources	25–35
Multi-lateral	
e.g., World Bank	20–25
Bi-lateral	
e.g., USAID, UKODA	10–15
Private sources	25–35
Others	15–20
Possible gap	0–10

(b) Official lending played the major role in all cases: about 75%–85% for electricity lending to low income countries and 55%–65% to the others, multilateral agencies providing about half of flows to low income countries.

(c) Official lending has significantly better financing terms, particularly average maturity periods which are two to three times those of private sources, 20–40 years versus 10–15 years.

(d) Electricity lending has taken about 15% of total borrowing by LDCs; if all electricity capital had been covered by borrowing this percentage would have been about 15%–18%; however, estimates of foreign exchange requirements for electricity 1976–1985 represent about 20% of their total foreign exchange requirements and any increase in this is deceptive.

Percentages of different kinds of capital making up electricity finance in the developing world in the 1990s are likely to be as shown in Table 8.32. For most developing countries there are consortia of donors, official and private, who meet once, possibly twice a year, usually chaired by a multilateral aid agency, to arrive at an aid package best suited to the country and the utility requirements, usually involving cofinancing (Chapter 5C).

8.6.4 *Guarantees*

Perhaps the greatest difficulty in utilities putting together financing plans lies in obtaining guarantees against socio-political changes, these often happening suddenly. Insurance against other forms of risk, technical, organisational and financial can be taken out by both donor and borrower more easily. Over the last ten years, the World Bank has put together a multilateral agency to cover socio-political and non-commercial risks (Chapter 5C), the Multilateral Investment Guarantee Agency (MIGA), consisting of a high proportion of Bank member countries who provide foreign investment in developing countries. MIGA provides long-term risk guarantees to investors against expropriation, war and civil disturbance losses, and provides advisory services. Coverage is available from MIGA for equity and other long-term investments by equity holders which are not made or irrevocably committed at the time of application for a MIGA guarantee, or similar non-commercial aspects.

8.7 Commentary summary

There is acceptance in the 1990s that developing countries cannot be treated by developed countries merely as juniors for the more advanced, or as primitives for the less developed. This in the past led to wrong approaches on the energy and electricity issues of: technology transfer; demand management; reliability standards; quality of supply; utility ownership and funding; regulation; investment programmes; and pricing. It has always been recognised that development needs energy, usually assuming more energy is needed per unit of development for developing countries compared with industrialised countries. In this matter something can be learned from studying the industrial histories of developed countries, especially the part played by energy. During the twentieth century electricity has obviously made an impression on development, and it will play an even larger role in the 1990s and beyond, although the sector, even in developing countries, must take account of recent changes in emphasis with respect to: demand management; increasing efficiency; conservation; environmental maintenance; private ownership and funding; competition; deregulation; time-of-use and dynamic pricing; consumer preference and reaction; short-term financial optimisation rather than long-term economic. There are special roles for government in developing countries under these changes.

There are certain government issues and options which developing countries must address for electricity:

(a) Methodologies for optimisation of development programmes, these being consistent throughout sectors
(b) Making best use of scarce national resources
(c) Consistent pricing rationale
(d) Efficient regulation
(e) Reconciling short-term financial utility objectives with long-term economic national objectives
(f) Best use of money, foreign and local, manpower and skills
(g) Training
(h) Appropriate, efficient, technology transfer
(i) Rural electrification, forming one component of rural energy development.

For many years developing countries have suffered from wrong technology transfer for electricity:

(a) Chasing indefinite gains in economies of scale
(b) Installing plant with high capital to running cost ratios
(c) No attending to geography, or lack of skilled labour, or cultural prejudices
(d) Lack of training programmes
(e) Choosing the 'wrong' plant, operationally, culturally, or strategically
(f) Giving attention only to plant successful in developed countries
(g) Too simple transfer of electricity utilisation plant.

There exist special aspects to financing electricity in developing countries:

(a) Obtaining right balances of foreign/local funding
(b) Ensuring each project has appropriate finance with respect to services, e.g. projects with long-term installation times obtaining long-term finance
(c) Obtaining enough local and foreign finance for efficient sector growth and operation, sometimes very difficult
(d) Setting financial targets and accounting rules for both publicly and privately owned utilities, within national economic energy plans, within also national and international accounting rules
(e) Ensuring debt servicing by correct pricing, and optimising private and public funding balances; in some developing countries electricity will lead in these matters, in others, government; or regulatory bodies will lead.

Traditionally electricity in developing countries has been publicly owned, regulated and controlled for planning, pricing and operation. In the 1990s private utility funding/ownership in developing more than in developed countries requires special regulation:

(a) Whilst changing over not losing development momentum nor national/foreign resources
(b) Obtaining the best balances of private/official/public capital
(c) Setting groundrules for debt service
(d) Ensuring regulatory bodies are independent of utilities and government, plus able to work properly.

Developing countries' debt service in the 1980s was astronomical, but public/private authorities will not allow this in the 1990s. Electricity is notoriously capital-intensive, and in developing countries is unlikely to be 'allowed' resources profligately as in the 1980s. This means changing plant types towards lower capital/operating cost ratios and charging realistic prices to cover debt servicing properly. Throughout the 1970s and 1980s, developing countries followed developed in optimising on the national economy. They will now follow developed countries towards private ownership and funding, etc., mentioned above towards optimising on the utility. However, unlike developed countries, developing countries must to a large measure still seriously consider utility planning against the national economy, and thus, long-term economic optimisation must still be done, seriously, frequently, and with conviction, probably by government. However, even in developing countries, electricity will be optimised greatly with respect to the utility, often giving different development programmes from those of the past:

(a) Diesels, gas turbines and combined cycle plants instead of large fossil fuel steam plants
(b) Smaller grids, or no grids
(c) Differently structured distribution and isolated systems. These have important bearings for rural electrification.

Short-term financial optimisation becoming the 'norm' in the 1990s has special significance for developing countries:

(a) Ensuring national economy/energy planning and pricing objectives are

not distorted by electricity policy, e.g. for local/foreign debt service, fuel substitutability, equity between consumers; uniform and fair financial targets
(b) Uniform accountancy rules
(c) Optimum use of scarce national capital, skills, labour, etc.
(d) Reconciling long-term economic with short-term financial aspects.

8.8 References and further reading

1 See World Bank Atlas and similar documents for 1990, 1991, etc., for data for particular countries, especially per capita income; also similar publications of the Inter-American Development Bank, Asian Development Bank, etc. and bilateral aid agencies, USAID, UKAID, Swedish Aid, etc. and UN agencies
2 *World Bank News*, April 1990, p. 5; also *World Bank News*, July 1990, pp. 1–3; also World Development Reports 1990, 1991, etc.
3 BERRIE, T.W. (1990) 'Can developing countries afford to be efficient in energy?, paper to the *Third World Energy Research Group* (Surrey University)
4 *Oxford Energy Forum*, **1**, No. 2 August 1990, pp. 3–8
5 HUMPHREY, W.S. and STANISLAV, J. (1979) 'Economic growth and energy consumption in the UK 1700 to 1975', *Energy Policy*, March, pp. 30–40, also, Dean, P. and Cole, W.A. (1969), 'British economic growth 1688 to 1959' (Cambridge University Press); also Nef, J.V. (1932) 'The rise of the British coal industry' (George Routledge and Sons Ltd); also, Feinstein, C. (1977), 'National income expenditure and output of the United Kingdom 1855–1965', Cambridge University Press; also Dean, P. (1968) 'New estimates of GNP for the United Kingdom', Review of income and wealth series, 14, No. 2; also, Mitchell, B. R. and Dean, P. (1977) 'Abstracts of British historical statistics' (Cambridge University Press); also 'UK digest of energy statistics', various years (HMSO, Holborn, London)
6 MUNASINGHE, M. (1989) 'Power for development; electricity in the Third World, *IEE Review*, March, pp. 101–105; also, World Bank (1988) 'A review of World Bank lending in electrical power', Energy series paper No. 2; also Moore, E.A. (1990) 'Electric power in developing countries in the 1990s' (World Bank, Washington DC USA)
7 SAUNDERS R.J. and JACHOUTEK, K. (1985) 'The electric power sector in developing countries', *Energy Policy*, August, pp. 320–325; also, Mackerron, G. (1987) 'Energy pricing criteria for developing countries versus practice in industrialised countries': paper for *Energy Pricing Workshop*, Study Group on Third World Energy Policy (Surrey University); also, Munasinghe, M. and Warford, J. (1981) 'Electricity pricing in developing countries' (The Johns Hopkins University Press, Baltimore, USA); also Munasinghe, M. and Schramm, G. (1980) 'Power—energy pricing case studies' (The World Bank, Washington DC)
8 DEHLMAN, C.J. (1989) 'Technological change in industry in developing countries', *Finance and Development*, June (Washington, DC) pp. 13–15; also, Moore, E. A. (1990) 'Electric power in developing countries in the 1990s' (The World Bank, Washington DC); also Friedman, E. (1978) 'External financing of power expansion in developing countries', Public Utilities Department, and updates with approximately the same title; also 'Cofinancing' World Bank pamphlet, (1980)

Epilogue

Even while this book was being written it was difficult to keep pace with the changes in the utilities sectors of most countries, the rate of change if anything accelerating and widening. Most of these changes are encapsulated in the relevant chapters and, where the way ahead was not certain, this has been plainly stated, the issues and options and likely outcome described. The required capital investment sums for electricity sectors worldwide are so large as to be meaningless, and few would dispute that for the 1990s official, government or public capital could never supply all that is necessary. Whatever the pros and cons of private versus public funding and ownership, everywhere the private sector must needs be tapped for a large proportion of the finance needed, and with finance supplied in any quantity there must come private ownership, at least in part of most electricity sectors, and again at least some regulation. Private finance will insist that short-term financial criteria are taken more seriously than possibly they have been in the past, but governments must always, sooner or later, in their turn, insist on the great importance of long-term economic criteria. Investment rules will therefore be mixed although, in a swing towards addressing the balance, the early 1990s will see a strong move towards the short-term financial. This will make an abrupt change in the type of generation ordered and a similar abrupt change in network structure of these electricity sectors still largely developing.

It seems likely that demand management will blossom worldwide once the message gets across that it is a cheaper alternative to new plant, and that dynamic, time-of-use, and spot tariffs will spread rapidly, even down to residential consumers. For similar reasons of saving money, efficiency improvements and conservation will be encouraged, even by utilities whose job it is to sell more electricity, not less. Efforts in environmental maintenance must eventually be decided mainly by economics and finance rather than by pressure groups, and at the level the world is prepared to pay for.

Developing countries' electricity sectors have always had special problems but, for the 1990s, another separate large group of countries has emerged which have similar difficulties, plus some difficulties of their own, namely the republics of the USSR and Eastern Europe. Whereas developing countries have had to show almost infinite patience in waiting for the special problems to be solved, it became obvious even by the early 1990s that this other group of countries would not follow suit. They will want solutions to be found during the years of the 1990s, even though some of their problems in electricity supply are similar to those of Western Europe and North America, which have not been properly solved to date!

The most urgent issues and options facing electricity sectors for the 1990s are:

(a) Regardless of utility ownership, how can enough monies be mustered to keep electricity sectors in a good sound state, technically, operationally, managerially, economically and financially?
(b) How can an optimum balance in the electricity sector be determined between government intervention, local authorities, regulators, utilities, consumers and the general public in any particular situation?
(c) Regardless again of ownership, how can electricity markets be made more efficient, e.g. by competition, dynamic pricing, buying electricity forward, futures electricity markets, agents and brokers?
Who should act as marketmakers, wholesale or retail?
(d) How can those responsible for decision-taking in electricity sectors take proper account of other sectors in the economy, especially rival sectors, e.g. gas, oil, coal?
(e) How will the nuclear debates be settled? Will nuclear fusion be developed in time to save world fuel starvation, or will some other energy source?

Index

Accounting costs
　see Financial costs
Accounting rules 121, 267, 268
　see also Financial targets
Acid rain 18, 106, 109, 110, 111, 116
　see also Nitrous oxide emissions
　　　　　Sulphur emissions
Agents for buying/selling 41, 131, 139, 177, 179, 180, 270
　see also Buying forward
　　　　　Futures markets
Aggregation approach
　see 'Bottom upwards'
Air conditioning 21, 57, 59, 67, 68, 70, 178
Annual rate of return on assets (xxvi), 45, 122, 140, 142, 148, 149, 170, 187, 188, 189, 194, 200
Autogeneration (xvii), (xxiii), (xxv), (xxvi), 43, 49, 62, 83, 133, 136–138, 175, 179, 183, 184, 186, 188, 195–196, 205
　see also Cogeneration

Biomass 8, 11, 14, 109
Blackouts (xx), 57, 74, 86, 131, 254
　see also Interruptible tariffs
　　　　　Outage costs
Boilers
　fluidised beds 90, 109
　super critical 182
Border prices (xvi), 37, 39, 156
　see also Efficiency prices
　　　　　International prices
　　　　　Opportunity costs
　　　　　World market prices
'Bottom upwards' 54–55
　see also Load forecasts
Brokers
　see Agents
Bulk supply tariff
　see Pricing, Bulk supply tariff
Buying forward 131, 134, 179, 204, 270
　see also Futures markets

'CALMS' 60, 63, 74
　see also Demand management, by physical control, by price
Capacity costs 158–162, 258
　see also Demand charge

Capital cost 29, 38, 56, 97, 137, 152, 176, 182–183, 188, 190, 200, 202–203, 207, 234, 249, 252, 258, 260, 264, 266
　see also Investment cost
Carnot cycle 182
CO_2 emissions 30–32, 108–116
　see also Global warming
　　　　　Greenhouse effect
Coal
　exploration 27
　extraction 10
　markets 10, 25, 35
　planning 28, 37, 52
　price 18, 24–25, 27, 39, 259
　　see also Energy pricing
　reserves 27
　supply/demand (xv), 9, 27
　usage 24–25
Cofinancing 38, 146
　see also Financial engineering
Cogeneration (xvii), (xxiii), (xxv), (xxvi), (xxviii), 43, 49, 55, 62, 83, 91, 133, 136–137, 175, 179, 183–184, 186, 188, 195–196, 203, 205
Combined cycle plant (xix), (xx), (xxvii), 86, 89–91, 109, 177, 182–183, 264, 267
Combined heat and power (xix), (xx), (xxiv), (xxvi), 86, 89–94, 109, 136, 204–205
　see also District heating
Common carrier 93, 127, 130–132, 192, 194
　see also Transmission, function
　　　　　Wheeling
Competition (xxvii), 27, 119, 122, 133, 140, 144, 146, 190, 266, 270
　see also Monopolies
　　　　　Pareto optimality
　between fuels (xvii), (xxi), 16
　distribution (xxiii), 119–121
　generation (xvii), (xxiii), 44, 49, 62, 123, 135–136, 139, 145, 177, 179, 190, 195
　new utilities (xxiii), (xxiv), 62, 126, 135–136, 140–141
　transmission (xxiii), 119–121
Computer models 29–30, 42, 51, 134–135, 156, 178, 186, 188
Consumer classes 57, 151, 162, 189, 213, 259
Consumer reaction
　electricity usage 105, 165, 266

price (xix), (xxvii), 30, 44–45, 56, 63–67, 162
supply standard (xix), (xxvii), 57, 64, 186
Consumer research
    see Load research
Co-operatives 104, 120, 121
Cost-benefit analysis (xxi), (xxii), 35, 107, 116, 119
    see also Least-cost solution
Cost-effectiveness 62, 89, 97, 104, 163
    see also Least-cost alternative choice
Cost minimisation
    see Least-cost solution
Current limiters (xviii)

Debt
    Third World (xxxi), 3, 120, 146, 234, 255–256, 267–268
Deforestation 19, 106, 256
Demand charge 169
    see also Capacity costs
Demand forecasting (xii), 28–29, 50, 56, 105, 128, 173, 179, 189, 201–202, 213–214, 254
    see also Load forecasts
Demand management (xx), (xxii), (xxvi), (xxvii), (xxviii), (xxx), 6–7, 17, 28, 41, 46, 48, 50, 57, 74, 79–84, 98, 101–106, 115, 130, 145, 164, 183, 184, 188, 190, 202, 205, 221, 249, 257, 260, 266, 269
    by contract (xix), 28, 62, 83, 176
    by physical control (xviii), (xix), 28, 48, 57–61, 71, 83, 165, 170
        see also 'CALMS'
                Mainsborne
                Pilot wire control
                Radio teleswitches
    by price (xix), 28, 48, 62–63, 83, 176, 179
        see also 'CALMS'
    by retrofit 28, 98
    success (xv), (xvii), (xviii), (xix), 48, 58
Depletion rent, fuels 19
Deregulation 175–180, 195, 258, 266
    see also Regulation
Desertification 106
Developed countries (xx), (xxv), (xxviii), (xxxi), 1–5, 7, 17, 19, 27, 30, 35, 37–42, 86, 97, 100, 106–107, 110, 123, 146, 172, 174, 191, 206, 208, 213, 221, 234, 249, 266
Developing countries (xv), (xvi), (xxii), (xxiv), (xxv), (xxvii), (xxix), (xxx), (xxxi), throughout Chapter 8
    conservation 99
    demand and reliability 55, 83, 84
    development programmes 188, 191, 196–197, 201–202, 206–207
    efficiency 91, 98
    electricity sector 38–42, 44

environmental maintenance 106–107, 110, 114–117
pricing 152, 164, 174
utilities ownership 120–123, 130, 137, 146–147
world outlook 1, 2, 5–9, 15, 16, 19, 29–36
Development programmes 105, 181, 260, 266
    see also Investment programmes
    distribution (xxvii), 42, 190, 196, 199, 202
    generation (xii), (xxvi), (xxviii), (xxx), 42, 181, 182
    optimisation (xxii), (xxv), (xxvi), (xxviii), (xxx), 28
    transmission (xxvii), (xxviii), 42, 189, 190
    utilisation 42, 202, 204–205, 266
Diesels (xxvii), 182–183, 196–197, 201–202, 260, 264, 267
Disaggregation 54
    see also Load forecasts
            'Top downwards'
Discounting (xii), 101, 156, 178, 188, 190, 259
    see also Present valuing
            Social time preference
            Time value of money
District heating 90, 136
    see also Combined heat and power
Dynamic pricing
    see Pricing, dynamic

Economic costs vs financial costs 30, 116, 151, 188
    see also Financial costs
            Shadow pricing
Economic efficiency 105, 108, 123, 139–140, 144–145, 151, 155, 159, 165, 189
Economic rent 10, 34
Economic return (xxii), (xxviii), 108, 116, 200–201
Economies of scale (xxiii), (xxxi), 10, 12, 15, 116, 119, 189, 206, 208–209, 251, 263–264, 266
Efficiency (xv), (xvi), (xxii), (xxvii), (xxx), 6–8, 17, 28–40, 46, 62, 76, 83, 86–91, 99–106, 111, 115, 120, 140, 145, 170, 181, 184, 189–190, 195, 202, 221, 234, 249–252, 259–260, 263, 269
    distribution (xx), 41, 83, 86–87, 95, 98, 254
    funding 97
    generation (xix), (xxiv), (xxv), 41, 57, 83, 86–91, 98, 116, 179, 182–183, 186
    markets 135, 146, 178, 195, 207
        see also Electricity, market efficiency
    prices (xxiv), 116
        see also Opportunity costs
                Border prices
    transmission (xx), 41, 83, 87, 95, 98, 254
    utilisation (xxviii), (xxix), 86, 96–97, 103, 254

Elasticity
  GDP growth to energy growth 6, 7, 15–19, 51, 99, 100, 234–235, 240–243, 247, 263, 266
  GDP growth to electricity growth 51–52, 100, 234, 249, 257, 263, 264
Electricity conservation (xxi), (xxvii), (xxx), 101, 103, 106, 130, 139, 170, 179, 184, 188, 190, 221, 269
  see also Energy conservation
  market composition 42–44, 48, 79, 136, 166, 175, 184
  market development 43, 48
  market shares (xvii), 42, 45–46, 55, 59, 178, 184, 259
Electricity, market efficiency 43–44, 48, 96, 99, 103, 106, 120–121, 165, 203, 259
  see also Efficiency, markets
  planning (xviii), 28, 37, 49, 50, 53, 56, 64, 98, 105, 156, 178, 186, 188–191
  sales (kWh) 54, 56, 63, 152, 190, 260, 264
    see also Peak demand (kW)
  sector reviews
    national (xv), (xvii), (xxx), 37, 41, 43, 45, 48, 53, 56, 107, 151, 183, 184
  supply-demand balance 184, 189, 190, 249
  utility (xviii), (xxx), 43, 45, 47–48, 53, 56
Energy
  audits 104–105
    see also Energy conservation
  conservation 6–8, 18, 28–40, 45, 62, 64, 83, 86, 89, 98, 100–105, 108, 111, 145, 160, 165, 181, 234, 249, 259–260
    see also Electricity conservation
         Energy, audits
  management 7–8, 19, 28–29, 35, 37, 40, 46
    see also Demand management
  planning (xv), 12–15, 28, 37, 39, 41
  pricing (xv), 8–9, 14, 28–36, 40–41, 54, 99, 102–103, 106–108, 156
    see also Electricity, price
         Coal, price
         Gas, price
         Oil, price
  sector reviews (xv), (xvi), 29, 40, 48
  supply-demand (xvi), 5, 28–29, 40, 52, 99
  usage (xvi), 28–30, 96
Environmental maintenance (xv), (xxi), (xxx), 7, 17, 21, 24, 28–30, 37, 40, 42, 46, 89, 106–110, 114–117, 165, 177, 182, 190, 221, 234, 259–260, 266, 269
  economics 30, 32, 110
  funding 107, 116–117
  generation 108
  responsibility (xxi)
Equity 38, 146, 151, 268
Extrapolation 50, 56
  see also Load forecasts

Finance, sources (xv), (xvi), (xxv), 42, 146–147, 149, 260, 263–265, 269
  see also Financial engineering
Financial costs 89, 123
  see also Economic costs
Financial engineering (xxv), 147–149, 186, 260, 267
  see also Cofinancing
       Finance, sources
       Financing plans
Financial return on investment 148–149, 255
  see also Annual rate of return on assets
Financial statements 186, 189
Financial targets 122, 141, 176, 189, 249, 267–268
  see also Accounting rules
Financing guarantees 38, 146, 149, 258–259, 265
  MIGA 147, 265, 266
Financing plans (xxv), 148–149, 186
  see also Finance, sources
Fuel
  calorific value 33
  mix 29
  substitution (xxi), (xxii), 6, 8, 16, 18–19, 25, 28, 31–34, 44, 54–55, 59, 64, 108, 116, 234, 268
Fuelwood (xv), 7–8, 28, 35, 37, 235, 237, 249, 256
  see also Renewables
Futures markets 41, 102–103, 131, 134, 139, 179, 270
  see also Buying forward
       Agents

Gas
  economic extraction 19–20
  exploration 20
  markets 19, 21, 23, 25
  planning 28, 37, 52
  price 1, 18–19, 23, 27, 39, 259
    see also Energy pricing
  reserves 19–24
  supply/demand balance (xv), 9, 19–21, 26
  turbines (xxvii), 160, 182–183, 196–197, 201–202, 260, 264, 267
Geothermal 8, 14, 109
Global warming 106, 108, 112–113
  see also CO_2 emissions
       'Greenhouse effect'
'Greenfield' situations 55, 191, 196, 202
'Greenhouse effect' 30, 32, 108, 110–113
  see also CO_2 emissions
       Global warming
       Nitrous oxide emissions

Hydro
  large 8–9, 12, 14, 108, 127, 152, 163, 182, 247
  micro 182, 196–198, 201–202, 235

International fuels 10–12, 21, 26–27, 156
International prices 97
  see also Border prices
Interruptible tariffs (xx), 21, 57, 64, 68, 170, 173–174, 178
  see also Blackouts
       Load shedding
Investment cost 38, 40, 81, 117, 234, 252, 254, 256, 259, 263
  see also Capital cost
Investment programmes 28, 135, 139, 151, 176, 258
  see also Development programmes

Least-cost solution (xxvi), (xxvii), (xxix), 43, 54, 56, 57, 103, 108, 139, 145, 156, 164–165, 184, 186, 188–190, 195–196, 199
  see also Cost-benefit analysis
       Cost-effective solution
Life expectancy 216–217, 231–233
Life-line tariffs 12
  see also Subsidised prices
Life-time return 116
LNG 21
Load factor 50, 183, 184
Load forecasts (xvii), (xxix), 28, 50, 53, 56, 199
  see also 'Bottom upwards'
       Demand forecasting
       Extrapolation
Load research 57, 97, 105–106
Load shedding (xix), 57, 68
  see also Interruptible tariffs
Long-run marginal cost (xii), (xxvi), 39, 43, 97, 155–158, 160, 165, 179, 188, 204, 259
LOLP 134, 135

Mainsborne control 58, 60
  see also Demand management by physical control
Marginal cost of production 10, 12–13, 27, 97, 102, 105
Marginal cost pricing
  see Pricing, marginal cost
Market makers (xxvii), 128, 132, 175–179, 194–195, 270
Market surveys 50, 55–56, 199
  see also Load forecasts
Merit order 127, 160–161, 194–195
Meters 161–162, 172
Monopolies 119–122, 125, 127, 130, 132, 138, 139, 166, 194
  see also Competition

National economic review (xviii), 39–40, 48–53, 81, 99, 107, 120, 139–140, 143, 151, 184, 207, 219, 267
National energy review 8, 28, 39, 40, 48, 50–51, 53, 99, 104, 107, 120, 143, 207, 267

Nitrous oxide emissions 108
  see also Acid rain
       Sulphur emissions
Non-OPEC oil 15, 17, 18
Non-renewable indigenous energy 8–12, 19, 37, 101, 156, 258, 260, 264
Nuclear 8, 18, 27, 32–35, 42, 62, 101, 108–112, 120, 128, 177, 182–183, 191, 270

Ocean current 8–9, 14
Off-peak demand 62–66, 156–159, 163–164, 203
  see also Peak demand
Oil
  economic production 15
  exploration 10, 11
  exporting countries (xvi), 37
  importing countries (xvi), 19, 23, 37, 256, 260
  planning 28, 52
  price (xv), (xvi), 1, 7–9, 15–19, 25, 27, 31, 34, 37, 39, 99, 100, 206, 249, 259–260
    see also Energy pricing
  reserves 20
  supply/demand balance (xv), 15–18, 240–242
Opportunity costs 29, 74, 106, 110, 156, 184, 259
  see also Border prices
       Efficiency prices
       Resource costs
Optimisation
  long-term economic (xxx), 41–43, 55, 57, 97–98, 101, 121, 179, 190, 204, 259, 266, 268, 269
  national economic (xxx), 28–29, 41, 155
  short-term financial (xxx), 41–42, 55, 57, 97–98, 101, 121, 179, 199, 259, 266, 268–269
  utilities 28, 41
Outage costs (xix), 74, 84, 178
  see also Blackouts
       Reliability of supply

Pareto optimality 44, 123
  see also Competition
Pay back period 198, 200
Peak demand (kW) 50, 54, 58, 62–64, 79–83, 95–96, 132–133, 152, 156–164, 184, 189–190, 193, 203
  see also Electricity sales (kWh)
       Off-peak demand
Pilot wire control 58, 60
  see also Demand management by physical control
Planning plant margin 83, 128, 179, 190, 262
  see also Risk
       Uncertainty
Plant-mix 160, 264

'Polluter pays' principle 108, 110–111, 114, 120
Pool purchasing price (PPP) 135
Pool selling price (PSP) 135
Population growth and energy (xv), (xxix), 1, 31, 34, 115–116, 213–218, 222, 234, 236–237
Premium fuels
  markets 23–24, 27, 100, 119, 156, 182
Present valuing 122
  see also Discounting
    Social time preference
    Time value of money
Pricing
  accounting cost (xxvi), 151, 156, 164
  bulk supply tariff 63, 132, 160, 162–163, 193, 195
  dynamic (xii), (xvii), (xx), (xxvi), (xxix), 43–44, 48, 74, 130, 132–133, 168, 179, 185, 188, 195–196, 203–205, 249, 257, 259, 266, 269–270
  economic efficiency 155, 157, 254, 266
  long-term contracts 133, 136–137
  marginal cost (xxvi), 97, 154, 156, 162–165
  prescribed (xxvi), 43, 56, 151, 164–165, 170, 186–187, 195, 203–204
  quality of supply premium (xxvi), 132, 166, 171, 178–179
  'spot' (xvii), (xx), (xxvi), (xxvii), 6, 43–44, 103, 135, 167–179, 188, 203–205, 269
  three part 153
  time of use (xvii), (xxvi), 43–44, 59, 62, 131, 156, 160–165, 168–171, 179, 203–204, 266, 269
  two part (xxvi), (xxvii), (xxix), 152, 162–165
Private funding (xv), (xxii), (xxiii), (xxiv), (xxv), (xxxi), 5, 17, 35, 38–39, 41, 43, 46, 48, 120–121, 127–128, 146, 179, 253, 263, 266, 269
Private ownership (xv), (xxii), (xxv), 1, 39, 41, 48, 54, 104, 119–121, 126–128, 131, 133, 137, 146, 149, 178–179, 183, 187, 195, 198, 201, 253, 258, 266–267, 269
  see also Privatisation
Privatisation (xvii), (xxii), (xxiii), (xxiv), (xxvii), 42–43, 54, 62, 94, 120–139, 143–145, 165, 190, 196, 258
  see also Private funding
    Private ownership
Probabilities (xii), 184
  see also Risk
Public funding (xxii), (xxiii), (xxiv), (xxv), (xxxi), 1, 38, 46, 128, 146, 234, 259, 266, 269
Public ownership (xxii), (xxiii), (xxv), 41, 43, 54, 104, 119, 121, 127, 132, 137, 141, 177–178, 183, 195, 258, 266–267, 269
Public utility commissions 139
  see also Regulation, responsibility

Radio teleswitches control 58, 60, 73
  see also Demand management by physical control
Regulation
  efficiency 43, 135, 137–138, 266
    see also Deregulation
  heavy (xvii), 43, 46–47, 119, 131, 138–140, 145, 177, 180
  legal (xxiv), (xxx), 133, 143–145
  light (xiii), (xvii), 47, 139–140, 144–145, 180, 194
  location (xxv), 47, 141
  monitoring (xxv), 138, 140, 143–145, 194
  responsibility (xvii), (xxiii), (xxiv), (xxx), 47, 49, 195
  role of authority (xvii), (xxiv), (xxx), 47, 49, 55, 141, 190, 259, 266, 267, 269
  transparency 140, 143–144, 178
Reliability of supply (xix), 46, 50, 57, 74–76, 84, 131, 138, 165, 181, 189, 190, 196, 198, 202–203, 251, 266
  see also Outage costs
    Planning plant margin
    Risk
Renewables (xxi), 12, 14–15, 18–19, 35, 41, 99, 179, 181
  see also Fuelwood
  indigenous fuels (xxviii), 8–9, 11–12, 37, 99, 207
Resource costs (xviii), 23, 39, 55–56, 107–108, 151, 234, 253, 258
  see also Opportunity costs
Revenue recovery 29, 45, 56, 110, 165, 176, 187, 189–190
'Ring fencing' 141
  see also Regulation
Rio Conference 38, 111
Ripple control
  see Mainsborne control
Risk (xii), 127, 133, 139, 144, 179, 180, 259
  see also Planning plant margin
    Probabilities
    Reliability of supply
Rural development 42
Rural electrification (xxx), 42, 104, 108, 116, 207, 266–267

Scarcity rent 12–13
Scenario planning 184, 188, 190, 200
Shadow pricing 108, 259
  see also Economic costs
Short-run marginal costs 179, 188
Small electricity systems 196
Social time preference 101
  see also Discounting
    Present valuing
    Time value of money
Solar energy 8, 11, 14, 181–182, 196
Space heating 21, 57–58, 79, 96

'Spot' pricing
  see Pricing, 'spot'
Steam raising 21, 23, 25, 27
Storage heaters
  see Space heaters
      Water heaters
Structural economic adjustment 136–137
Subsidised prices 30, 42, 62
  see also Life-line tariffs
Sulphur emissions 30, 114
  see also Acid rain
      Nitrous oxide emissions
      'Polluter pays'
System marginal price (SMP) 134–135

Technology transfer 206, 213, 260, 266
  appropriate 40, 206–207, 266
  expatriate role 210
  innovation 54, 204, 207–208
  patents 211
  process (xxii), (xxx), 42, 116, 206, 266
  training (xxx), 42, 206, 209–210, 266
Tidal energy 8, 9, 14, 181
Time of use pricing
  see Pricing, time of use
Time value of money 201
  see also Discounting
      Present valuing
'Top downwards' 54, 207
  see also Disaggregation
Toronto Conference 30–32
Transmission
  distribution owned 132
  function 192–193
  generation owned 133
  market maker 131
    see also Market makers
  pooling
  purchasing agent 132

Transparency
  see Regulation, transparency
Trend forecasts 55–56
'Trickle down'
  see 'Top down'

UK energy development 238–248
Uncertainty 75–77, 83, 259
  see also Planning plant margin
      Reliability
      Risk
Uranium 9

Value of lost load (VOLL) 134–135

Water heaters 58–59, 67–68, 79, 96
  see Space heaters
Water power
  see Hydro, micro
Wave energy 8, 14, 109
Wheeling 175, 181, 192–196
  see also Common carrier
Willingness to pay 81, 151, 204
  see also Consumer reaction
Wind energy 8, 11–12, 14, 109, 177, 181, 196, 235
World Bank (xxii), (xxiii), (xxix), (xxx), 39, 40, 95–96, 107, 116, 146–147, 207, 211, 213, 216–217, 219–224, 249–254, 257–265
World economic growth (xv), 16, 18, 34–35, 39, 53, 99, 115, 211–213, 219–230, 237–238, 257, 260
World energy growth (xv), 1, 7, 20, 30, 35, 39, 53, 99, 115, 234, 240
World market prices 106
  see also Border prices